Linear Algebra

Linear Algebra

A. MARY TROPPER

Senior Lecturer in Mathematics
Queen Mary College, University of London

NELSON

THOMAS NELSON AND SONS LTD

36 Park Street London W1
P.O. Box 336 Apapa Lagos
P.O. Box 25012 Nairobi
P.O. Box 21149 Dar es Salaam
P.O. Box 2187 Accra
77 Coffee Street San Fernando Trinidad

THOMAS NELSON (AUSTRALIA) LTD
597 Little Collins Street Melbourne 3000

THOMAS NELSON AND SONS (SOUTH AFRICA) (PROPRIETARY) LTD
51 Commissioner Street Johannesburg

THOMAS NELSON AND SONS (CANADA) LTD
81 Curlew Drive Don Mills Ontario

THOMAS NELSON AND SONS
Copewood and Davis Streets Camden 3 N.J. 08103

First published in Great Britain 1969

17 761005 0 (boards)

17 771010 1 (paper)

Made and printed in Great Britain at the Pitman Press, Bath

Contents

Preface

Linear algebra is becoming a more and more vital tool for engineers, scientists, and economists. Systems theory and many other modern topics of applied mathematics freely make use of algebraic techniques. There are many very excellent books on the subject, but they are mainly for mathematicians and usually assume considerable familiarity with abstract and axiomatic methods; they are therefore not suitable as a first introduction for non-specialists. This book is an attempt to present the theory so that it can be understood by readers who are not primarily mathematicians. Nevertheless it might be of value to those mathematics students who find the transition from school to university very difficult. Many examples are worked in the text, both as motivation and as illustration of the theory, and problems for the reader are given at the end of each chapter, with their answers at the end of the book.

It is hoped that anyone who has read and understood this book will be able to read much of the standard literature that is readily available. Some suitable books are listed at the end; many of them contain most of the material of the present book and go more deeply into the subject.

A.M.T.

Preface

1 · *Groups and Fields*

Modern algebra is concerned with the structure of mathematical systems. Many of the structures appear over and over again in the systems we meet, not only in pure mathematics, but also in applications to physical science. For this reason we study the structures themselves and we are then able to apply the results of our study to a wide variety of problems and achieve a considerable unification and economy of effort.

1.1 A permutation group

Consider the problem of arranging three elements in order. Denoting the elements by 1, 2, 3, one possible order is 123. There are six possible arrangements and all can be obtained by *permutations* of this one. Let us denote by P the permutation that replaces 1 by 2, 2 by 3, and 3 by 1. Thus P transforms (123) into (231). The transformation that leaves each element unchanged is called the identity and denoted by I. We define the six possible permutations as follows:

$$I \quad (123) \to (123) \qquad R \quad (123) \to (132)$$
$$P \quad (123) \to (231) \qquad S \quad (123) \to (321)$$
$$Q \quad (123) \to (312) \qquad T \quad (123) \to (213)$$

Note that P can also be written as $(231) \to (312)$, etc.

Now define the product $X \circ Y$ of two permutations as the result of applying *first* the permutation Y and *then* the permutation X. The product is clearly also a permutation. For example

$$S \text{ is } (123) \to (321)$$

and $$R \text{ is } (321) \to (231)$$

so that $$R \circ S \text{ is } (123) \to (231).$$

However, this is P and we have $R \circ S = P$. On the other hand, $S \circ R$ is $(123) \to (312)$. We see this by noting that R sends 1 into 1, S sends 1 into 3, so that $S \circ R$ sends 1 into 3, etc. Thus $S \circ R = Q$ and $S \circ R \neq R \circ S$. We now construct a multiplication table for these six permutations.

	I	P	Q	R	S	T
I	I	P	Q	R	S	T
P	P	Q	I	T	R	S
Q	Q	I	P	S	T	R
R	R	S	T	I	P	Q
S	S	T	R	Q	I	P
T	T	R	S	P	Q	I

The product $R \circ S$ appears in the row opposite the entry R in the left-hand column and in the column below the entry S in the top row. Certain properties of the array of products in this multiplication table are immediately obvious. Every row contains each of the elements I, P, Q, R, S, T once, and once only. The same is true of every column. Thus each row and each column is a permutation of $IPQRST$. If X is any one of the six elements, $X \circ I = I \circ X = X$, and we say that I is an *identity* element. Further, since every row contains exactly one entry I, it follows that for any element X in the set there exists a unique element Y in the set such that $X \circ Y = I$. Since the I's are symmetrically placed with respect to the leading diagonal (from top left-hand corner to bottom right-hand corner) it then follows that $Y \circ X = I$. We regard Y as the inverse of X. There is one more property that is not immediately obvious from the table. Using the table we see that

$$P \circ (Q \circ R) = P \circ S = R,$$

$$(P \circ Q) \circ R = I \circ R = R,$$

so that

$$P \circ (Q \circ R) = (P \circ Q) \circ R.$$

We can easily verify that this is generally true, i.e.,

$$X \circ (Y \circ Z) = (X \circ Y) \circ Z$$

for any three permutations X, Y, Z belonging to the set. We say that multiplication is *associative*.

An operation which associates with any pair of elements X and Y (in that order) in a set, a unique element $X \circ Y$ in the set, is called a *closed binary operation* on the set (*closed* because the 'product' remains in the set).

1.2 Definition of a group

A set of elements G together with a closed binary operation on the set is called a *group* if the following three conditions are satisfied:

(1) the operation is associative, i.e., $x \circ (y \circ z) = (x \circ y) \circ z$ for all elements x, y, z in G;

(2) the set contains an identity element e such that $x \circ e = e \circ x = x$ for all x in G;

(3) corresponding to every element x in G there exists an element x^{-1} in G (called the inverse of x) such that $x \circ x^{-1} = x^{-1} \circ x = e$.

It is easy to see that the identity element is unique. For suppose that e satisfies (2) and (3) and that $x \circ f = f \circ x = x$ for all x in G. Then

$$x^{-1} \circ (x \circ f) = x^{-1} \circ x = e$$

and also $$x^{-1} \circ (x \circ f) = (x^{-1} \circ x) \circ f = e \circ f = f.$$

Thus $$e = f.$$

Again, the inverse x^{-1} of any element x is unique. For if

$$x^{-1} \circ x = e = y \circ x,$$

then

$$(x^{-1} \circ x) \circ x^{-1} = (y \circ x) \circ x^{-1},$$

$$x^{-1} \circ (x \circ x^{-1}) = y \circ (x \circ x^{-1}),$$

$$x^{-1} \circ e = y \circ e,$$

$$x^{-1} = y.$$

The permutation group discussed in Section 1.1 clearly satisfies all the above conditions and the identity element is I. It is called the *symmetric group of degree 3*. We see from the table that $P^{-1} = Q$, $Q^{-1} = P$, $R^{-1} = R$, $S^{-1} = S$, $T^{-1} = T$.

1.3 Further examples of groups

The symmetric group of degree 3 is an example of a group of transformations. Another example is the group of symmetries of an equilateral triangle. These are the motions of the triangle that bring it into coincidence with itself. Let the triangle be ABC, lettered in a counter-clockwise sense, and let I be the transformation that leaves its position unchanged. Let P, Q be the clockwise rotations of the triangle through $120°$, $240°$ respectively. Let R, S, T be the rotations through $180°$ about the altitudes through A, B, C respectively. Defining the product of two transformations in the obvious way, the reader should construct a multiplication table for the six transformations I, P, Q, R, S, T. It is exactly the same table as for the previous example. The group of symmetries of the triangle is *isomorphic* with the symmetric group of degree 3.

A correspondence between two sets of elements S_1 and S_2 is said to be one–one if to every element x_1 in S_1 there corresponds *exactly* one element x_2 in S_2 and vice versa. A one–one correspondence between two groups G_1 and G_2, which is such that $x_1 \circ y_1$ corresponds to $x_2 \circ y_2$ whenever x_1, y_1 correspond to x_2, y_2 respectively, is called an *isomorphism* between the two groups. Every result that is true for G_1 is also true for G_2, with a suitable interpretation of the binary operation in each case.

The set of integers forms a group when the binary operation is ordinary addition. For, if a, b, c are any integers, $a + b$ is another integer and

(1) $(a + b) + c = a + (b + c)$,
(2) $a + 0 = 0 + a = a$,
(3) $a + (-a) = (-a) + a = 0$.

The integer zero is the identity element in this group which, unlike the two previous examples, has infinitely many elements. The group has the additional property that $a + b = b + a$ for all integers a and b. Such a group (i.e., a group for which $x \circ y = y \circ x$ for all x and y) is said to be *commutative* or *abelian*.

The set of all real numbers is an abelian group under addition and so is the set of all complex numbers. The identity element in the additive group of complex numbers is $0 + 0i$, and the inverse of $a + ib$ is $-a - ib$. The set of all non-zero complex numbers is an abelian group under multiplication. The identity element is $1 + 0i$ and the inverse of $a + ib$, where $a^2 + b^2 \neq 0$, is $(a - ib)/(a^2 + b^2)$.

Here is one more example of a finite group. We divide the integers into three classes. All integers of the form $3n$, where n is an integer, go into one class which we call 0. Thus all the numbers $\ldots -9, -6, -3, 0, 3, 6, 9, \ldots$ are identified with 0. Again, all integers $3n + 1$, i.e., $\ldots -8, -5, -2, 1, 4, 7, \ldots$ form the class 1 and all the integers $3n + 2$, i.e., $-7, -4, -1, 2, 5, 8, \ldots$, form the class 2. The numbers 0, 1, 2 are simply the remainders, or *residues*, when the elements of the classes are divided by 3. We call the classes *residue classes modulo 3* and we write

$$\begin{aligned} 3n &= 0 \pmod{3}, \\ 3n + 1 &= 1 \pmod{3}, \\ 3n + 2 &= 2 \pmod{3}. \end{aligned}$$

If $x = 0$ and $y = 0 \pmod{3}$, then clearly $x + y = 0 \pmod{3}$. Thus the sum of any two elements in the class 0 is also in the class 0, and we write this symbolically as $0 + 0 = 0$. If $x = 1$ and $y = 2$, then $x + y = 0 \pmod{3}$. Again, we write this $1 + 2 = 0$. We can construct an addition table for the residue classes as follows:

+	0	1	2
0	0	1	2
1	1	2	0
2	2	0	1

Thus we see that the residue classes form an abelian group under addition, the identity element being 0. The multiplication table is as follows:

×	0	1	2
0	0	0	0
1	0	1	2
2	0	2	1

If we exclude the zero, obtaining the multiplication table

×	1	2
1	1	2
2	2	1

we see that the remaining classes 1 and 2 form an abelian group under multiplication with identity element 1.

1.4 Fields

We saw in Section 1.3 that the complex numbers form an abelian group under addition and that, if we exclude zero, they also form an abelian group under multiplication. They are a very important example of a *field*. We frequently meet sets of elements on which two distinct operations are defined. These operations are often, but not necessarily, addition and multiplication. We usually call the additive identity zero and the multiplicative identity unity (or the unit element). Note that the real numbers also form an additive abelian group and that the non-zero real numbers form a multiplicative abelian group. Moreover, the following conditions are satisfied:

$$a \times 0 = 0 \times a = 0 \text{ for all } a,$$
$$a(b + c) = ab + ac, \qquad (a + b)c = ac + bc \text{ for all } a, b, c.$$

We say that addition is *distributive* under multiplication. Every element a has an additive inverse $-a$ and every element $a \neq 0$ has a multiplicative inverse a^{-1}. We now define a field.

A *field* is a set F of elements closed under two binary operations which we shall call addition and multiplication, and satisfying the following conditions:

(1) under addition F is an abelian group with identity 0;
(2) under multiplication the elements of F excluding 0 form an abelian group;
(3) addition is distributive under multiplication.

We can deduce the important *cancellation law* for a field. If $ab = ac$ and $a \neq 0$, then $b = c$. For if $a \neq 0$, a^{-1} exists. Then

$$a^{-1}(ab) = a^{-1}(ac),$$
$$(a^{-1}a)b = (a^{-1}a)c,$$
$$b = c.$$

Similarly if $ba = ca$ and $a \neq 0$, then $b = c$.

In any field F the equation $a + x = b$, where a and b are in F, always has a unique solution $x = b + (-a)$ which is written $b - a$. If $a \neq 0$, the equation $ax = b$ has the unique solution $x = a^{-1}b$.

1.5 Some important fields

We have already seen that the real numbers form a field which we usually denote by $\boldsymbol{R}$. We have also seen in Section 1.3 that the complex numbers satisfy the conditions for a field and we denote this by $\boldsymbol{C}$. In the same paragraph we saw that the set $\boldsymbol{Z}_3$ of integers modulo 3 is a field. It can be shown that, if p is any prime number, the set $\boldsymbol{Z}_p$ of integers modulo p is a field. It is instructive to construct addition and multiplication tables for $\boldsymbol{Z}_6$ and show that this is *not* a field.

The sum and product of any two rational numbers are both rational numbers, and the set $\boldsymbol{Q}$ of all rational numbers is a field with additive identity 0 and multiplicative identity 1.

Henceforth, instead of writing 'x is a real number', or 'x belongs to $\boldsymbol{R}$', we shall use the generally accepted abbreviation $x \in \boldsymbol{R}$. The symbol $\in$ means 'belongs to' or 'is an element of'. Similarly we shall write $a \in \boldsymbol{C}$, $b \in F$ and so on. In the next chapter we shall use the concept of a field in our definition of a *vector space*, the primary object of our study.

Problems

Prove that the following sets are groups and find the identity element in each case.

1.1 The set of all rotations about a fixed axis.

1.2 The set of all translations in a plane.

1.3 The numbers 1, ω, ω^2 under multiplication, where $\omega^3 = 1$, $\omega \neq 1$. (Note that $\omega^2 + \omega + 1 = 0$.)

1.4 Numbers of the form $a + b\sqrt{2}$, where $a, b \in \boldsymbol{Q}$, under addition.

1.5 The set of 2×2 matrices under addition.

1.6 The set of non-singular 2×2 matrices under multiplication.

Prove that the following sets are fields and find the zero and unit elements in each case, the operations being addition and multiplication.

1.7 Numbers of the form $a + b\sqrt{2}$, where $a, b \in \boldsymbol{Q}$.

1.8 Numbers of the form $a + b\omega$, where $a, b \in \boldsymbol{Q}$, $\omega^3 = 1$, $\omega \neq 1$.

1.9 The integers modulo 5.

1.10 All functions of the form $P(x)/Q(x)$, where $P(x)$ and $Q(x)$ are polynomials in x with real coefficients and $Q(x) \neq 0$.

1.11 Is the set of all non-singular matrices of order 2×2 a field? If not, why not?

2 · *Vector Spaces*

The concept of a vector space is a generalization of the familiar ordinary space of three-dimensional vectors. We have a set of elements $\boldsymbol{u}, \boldsymbol{v}, \boldsymbol{w}, \ldots$ called vectors which can be added (by the parallelogram law) so that $\boldsymbol{u} + \boldsymbol{v} = \boldsymbol{v} + \boldsymbol{u}$ is another vector and $(\boldsymbol{u} + \boldsymbol{v}) + \boldsymbol{w} = \boldsymbol{u} + (\boldsymbol{v} + \boldsymbol{w})$ for all vectors $\boldsymbol{u}, \boldsymbol{v}, \boldsymbol{w}$. The vector $\boldsymbol{0}$ (having zero magnitude) is such that $\boldsymbol{u} + \boldsymbol{0} = \boldsymbol{0} + \boldsymbol{u} = \boldsymbol{u}$ for every vector $\boldsymbol{u}$, and the vector $-\boldsymbol{u}$ (equal in magnitude to $\boldsymbol{u}$ but opposite in direction) is such that

$$\boldsymbol{u} + (-\boldsymbol{u}) = (-\boldsymbol{u}) + \boldsymbol{u} = \boldsymbol{0}.$$

Thus this set of vectors forms an abelian group under addition. Again, if x, y are any real numbers, the vectors $x\boldsymbol{u}$, $y\boldsymbol{u}$ are defined and we have

$$x(\boldsymbol{u} + \boldsymbol{v}) = x\boldsymbol{u} + x\boldsymbol{v},$$

$$(x + y)\boldsymbol{u} = x\boldsymbol{u} + y\boldsymbol{u},$$

$$(xy)\boldsymbol{u} = x(y\boldsymbol{u}).$$

The set of vectors forms a *vector space* over the field of real numbers.

2.1 Definition of a vector space

We now generalize by taking $\boldsymbol{u}, \boldsymbol{v}, \boldsymbol{w}, \ldots$ to be elements of an additive abelian group V and $x, y, \ldots$ to be elements of any field F, provided that there exists a method of combining any element $x \in F$ with any element $\boldsymbol{v} \in V$ to give a unique element $x\boldsymbol{v} \in V$. This is known as *scalar multiplication*. We give the following formal definition.

Let F be a given field and V a given additive abelian group. Let a scalar multiplication be defined which combines any element $x \in F$ with any element $\boldsymbol{v} \in V$ to give a unique element $x\boldsymbol{v} \in V$ such that

(1) $x(\boldsymbol{u} + \boldsymbol{v}) = x\boldsymbol{u} + x\boldsymbol{v}$, $(x + y)\boldsymbol{u} = x\boldsymbol{u} + y\boldsymbol{u}$,
(2) $(xy)\boldsymbol{v} = x(y\boldsymbol{v})$, $1\boldsymbol{v} = \boldsymbol{v}$,

for all $\boldsymbol{u}, \boldsymbol{v} \in V$ and all $x, y \in F$. 1 is the unit element in F. V is then called a *vector space over the field* F, sometimes written (V, F). The elements of V are called *vectors*, by analogy with the geometrical example given above, and the elements of F are called *scalars*. Vectors will be printed in boldface type to distinguish them from scalars. They are not necessarily geometrical in character and we shall now give some examples to show how very different from one another vector spaces can be, even though they all have the same mathematical structure.

2.2 Examples of vector spaces

Example 1. Let V be the set of all 2×2 matrices with real elements. Thus $A = \begin{bmatrix} a_{11} & a_{12} \\ a_{21} & a_{22} \end{bmatrix} \in V$ when a_{11}, a_{12}, a_{21}, a_{22} are all real. If $B = \begin{bmatrix} b_{11} & b_{12} \\ b_{21} & b_{22} \end{bmatrix} \in V$, then

$$A + B = \begin{bmatrix} a_{11} + b_{11} & a_{12} + b_{12} \\ a_{21} + b_{21} & a_{22} + b_{22} \end{bmatrix} = B + A \in V.$$

The matrix $0 = \begin{bmatrix} 0 & 0 \\ 0 & 0 \end{bmatrix}$ is an additive identity, and each matrix A has an additive inverse $-A = \begin{bmatrix} -a_{11} & -a_{12} \\ -a_{21} & -a_{22} \end{bmatrix} \in V$. Thus V is an additive abelian group. If x, y are any real numbers,

$$xA = \begin{bmatrix} xa_{11} & xa_{12} \\ xa_{21} & xa_{22} \end{bmatrix} \in V$$

and we have

$$x(A + B) = xA + xB, \qquad (x + y)A = xA + yA,$$
$$(xy)A = x(yA), \qquad 1A = A.$$

We see, therefore, that V is a vector space over the field $\boldsymbol{R}$ of real numbers.

Example 2. Let V be the set of all numbers of the form $a + b\sqrt{2} + c\sqrt{3}$, where $a, b, c \in \boldsymbol{Q}$. Clearly, V is an additive abelian group. The identity (or zero) element is $0 + 0\sqrt{2} + 0\sqrt{3}$ and the inverse of $a + b\sqrt{2} + c\sqrt{3}$ is $-a - b\sqrt{2} - c\sqrt{3}$. If x is any rational number, $x(a + b\sqrt{2} + c\sqrt{3}) = xa + xb\sqrt{2} + xc\sqrt{3}$ also belongs to V, and the conditions (1) and (2) of Section 2.1 are easily seen to be satisfied. Thus V forms a vector space over $\boldsymbol{Q}$.

Example 3. Let F be a given field and $V_n(F)$, where n is a positive integer, be the set of all n-tuples $(x_1, x_2, \ldots, x_n)$ consisting of n elements of F in a specified order. We define addition and scalar multiplication as follows:

$$(x_1, x_2, \ldots, x_n) + (y_1, y_2, \ldots, y_n) = (x_1 + y_1, x_2 + y_2, \ldots, x_n + y_n),$$

$$k(x_1, x_2, \ldots, x_n) = (kx_1, kx_2, \ldots, kx_n), \text{ where } k \in F.$$

The reader should verify that $V_n(F)$ is a vector space over F. We often use the shorter notation F_n for $V_n(F)$. In particular $V_n(\boldsymbol{R})$ and $V_n(\boldsymbol{C})$ are denoted by $\boldsymbol{R}_n$ and $\boldsymbol{C}_n$.

The vector spaces of examples 1 and 2 both contain infinitely many elements and so does that of example 3 if the field F is infinite. However, suppose that F is the field $\boldsymbol{Z}_2$ of integers modulo 2. This has two elements, 0 and 1, and V therefore contains only four elements. They are (0, 0), (0, 1), (1, 0), (1, 1).

2.3 Subspaces of a vector space

Let V be a vector space over F and let M be a subset of V. If M is also a vector space over F (with the same rules for addition and scalar multiplication) we call it a *subspace* of V. Clearly every subspace M of V must contain the zero vector $\boldsymbol{0}$ and the set consisting of the zero vector alone, written $\{\boldsymbol{0}\}$, is a subspace. If we can show that, for all $\boldsymbol{u}, \boldsymbol{v} \in M$ and all $x \in F$, $\boldsymbol{u} + \boldsymbol{v} \in M$ and $x\boldsymbol{u} \in M$, then M is a subspace. For, taking $x = -1$ (the additive inverse of the unit element 1 in F) we have $-\boldsymbol{u} \in M$,† and hence $\boldsymbol{u} - \boldsymbol{u} = \boldsymbol{0} \in M$. Thus M contains a zero element and if $\boldsymbol{u} \in M$, its inverse $-\boldsymbol{u} \in M$ also. The associative and distributive laws are automatically satisfied in (M, F) since they are satisfied in (V, F).

The *intersection* of two subspaces M_1 and M_2, written $M_1 \cap M_2$, is defined to be the set of elements common to M_1 and M_2. We now show

† See Problem 2.9.

that $M_1 \cap M_2$ is also a subspace of V. Let $\boldsymbol{u}, \boldsymbol{v} \in M_1 \cap M_2$. Then $\boldsymbol{u}, \boldsymbol{v} \in M_1$, so that $\boldsymbol{u} + \boldsymbol{v} \in M_1$ (M_1 being a subspace). Similarly $\boldsymbol{u}, \boldsymbol{v} \in M_2$, so that $\boldsymbol{u} + \boldsymbol{v} \in M_2$. Hence $\boldsymbol{u} + \boldsymbol{v} \in M_1 \cap M_2$. Again $x\boldsymbol{u} \in M_1$ and $x\boldsymbol{u} \in M_2$ for all $x \in F$. Hence $x\boldsymbol{u} \in M_1 \cap M_2$, and the two conditions for a subspace are satisfied.

2.4 Spanning sets

V is a vector space over F and S is a set of vectors $\boldsymbol{v}_1, \boldsymbol{v}_2, \ldots, \boldsymbol{v}_k \in V$. Let M be the set of all vectors of the form $x_1\boldsymbol{v}_1 + x_2\boldsymbol{v}_2 + \cdots x_k\boldsymbol{v}_k$, where $x_1, x_2, \ldots, x_k \in F$. Now M is a subset of V; moreover, the sum of any two vectors in M belongs to M and so does a scalar multiple of any vector in M. M is therefore a subspace of V and we call it the *subspace spanned by* S and write it $M = [\boldsymbol{v}_1, \boldsymbol{v}_2, \ldots, \boldsymbol{v}_k]$. The set of vectors S is called a *spanning set* for M.

As an illustration, consider the example of the 'ordinary' space of three-dimensional vectors. It is well known that if $\boldsymbol{i}$, $\boldsymbol{j}$, $\boldsymbol{k}$ are three given mutually perpendicular vectors, then any vector $\boldsymbol{v}$ can be written uniquely in the form $\alpha\boldsymbol{i} + \beta\boldsymbol{j} + \gamma\boldsymbol{k}$, where $\alpha, \beta, \gamma \in \boldsymbol{R}$, and $\boldsymbol{i}, \boldsymbol{j}, \boldsymbol{k}$ is a spanning set for the space. The subspace spanned by $\boldsymbol{i}$ and $\boldsymbol{j}$ (consisting of all vectors of the form $x\boldsymbol{i} + y\boldsymbol{j}$) is the set of all vectors in the plane of $\boldsymbol{i}$ and $\boldsymbol{j}$. More generally, if $\boldsymbol{u}$, $\boldsymbol{v}$, $\boldsymbol{w}$ are three distinct non-zero and non-coplanar vectors, then any vector $\boldsymbol{r}$ can be written uniquely in the form $\alpha\boldsymbol{u} + \beta\boldsymbol{v} + \gamma\boldsymbol{w}$, and $\boldsymbol{u}$, $\boldsymbol{v}$, $\boldsymbol{w}$ is a spanning set for the space. Moreover, no *two* vectors could be a spanning set for the whole space, since $\alpha\boldsymbol{u} + \beta\boldsymbol{v}$ always lies in the plane of $\boldsymbol{u}$ and $\boldsymbol{v}$. Thus every spanning set contains at least *three* elements, and any spanning set containing just three elements is called a *basis*. We now extend this idea to vector spaces in general.

2.5 Linearly independent sets and bases

An expression of the form $x_1\boldsymbol{v}_1 + x_2\boldsymbol{v}_2 + \cdots + x_k\boldsymbol{v}_k$, where $\boldsymbol{v}_1, \boldsymbol{v}_2, \ldots, \boldsymbol{v}_k$ are vectors and $x_1, x_2, \ldots, x_k$ are scalars in F, is called a *linear combination* of the vectors $\boldsymbol{v}_1, \boldsymbol{v}_2, \ldots, \boldsymbol{v}_k$. If the x's are not all zero it is called a *non-trivial* linear combination, but if the x's are all zero it is called *trivial*. We say that the vectors $\boldsymbol{v}_1, \boldsymbol{v}_2, \ldots, \boldsymbol{v}_k$ are *linearly dependent* if there exists a non-trivial linear combination of them equal to the zero vector $\boldsymbol{0}$. Thus

$$x_1\boldsymbol{v}_1 + x_2\boldsymbol{v}_2 + \cdots + x_k\boldsymbol{v}_k = \boldsymbol{0},$$

where $x_i \neq 0$ for at least one i.

On the other hand, the vectors $\boldsymbol{v}_1, \boldsymbol{v}_2, \ldots, \boldsymbol{v}_k$ are said to be *linearly independent* if the only linear combination of them equal to $\boldsymbol{0}$ is the trivial one. In this case $x_1\boldsymbol{v}_1 + x_2\boldsymbol{v}_2 + \cdots + x_k\boldsymbol{v}_k = \boldsymbol{0}$ if and only if $x_1 = x_2 = \cdots = x_k = 0$. A single non-zero vector $\boldsymbol{v}$ is necessarily independent, since $x\boldsymbol{v} = \boldsymbol{0}$ if and only if $x = 0$.

THEOREM 2.1. The non-zero vectors $\boldsymbol{v}_1, \boldsymbol{v}_2, \ldots, \boldsymbol{v}_n \in V$ are linearly dependent if and only if one of the vectors $\boldsymbol{v}_k$ is a linear combination of the preceding ones $\boldsymbol{v}_1, \boldsymbol{v}_2, \ldots, \boldsymbol{v}_{k-1}$.

Proof. If $\boldsymbol{v}_k = x_1\boldsymbol{v}_1 + x_2\boldsymbol{v}_2 + \cdots x_{k-1}\boldsymbol{v}_{k-1}$, then

$$x_1\boldsymbol{v}_1 + \cdots + x_{k-1}\boldsymbol{v}_{k-1} + (-1)\boldsymbol{v}_k + 0\boldsymbol{v}_{k+1} + \cdots + 0\boldsymbol{v}_n = \boldsymbol{0}$$

and the vectors $\boldsymbol{v}_1, \boldsymbol{v}_2, \ldots, \boldsymbol{v}_n$ are linearly dependent. Conversely, if the vectors are linearly dependent, then

$$x_1\boldsymbol{v}_1 + x_2\boldsymbol{v}_2 + \cdots + x_n\boldsymbol{v}_n = \boldsymbol{0},$$

where the x_i are not all zero. Let k be the largest value of i for which $x_i \neq 0$. Now if $k = 1$ this implies $x_1\boldsymbol{v}_1 = \boldsymbol{0}$ with $\boldsymbol{v}_1 \neq \boldsymbol{0}$, so that $x_1 = 0$, giving a contradiction. Hence $k > 1$ and we may write

$$\boldsymbol{v}_k = -(x_1/x_k)\boldsymbol{v}_1 - (x_2/x_k)\boldsymbol{v}_2 - \cdots - (x_{k-1}/x_k)\boldsymbol{v}_{k-1},$$

giving $\boldsymbol{v}_k$ as a linear combination of $\boldsymbol{v}_1, \boldsymbol{v}_2, \ldots, \boldsymbol{v}_{k-1}$.

A *linearly independent spanning set* $\boldsymbol{v}_1, \boldsymbol{v}_2, \ldots, \boldsymbol{v}_n$ of a vector space V is called a *basis* of V. Unless $V = \{\boldsymbol{0}\}$ the basis vectors are non-zero, since we can eliminate any zero vector from a spanning set. In the example discussed in Section 2.4 the vectors $\boldsymbol{i}, \boldsymbol{j}, \boldsymbol{k}$ form a basis of the space.

THEOREM 2.2. Any vector space which has a finite spanning set contains a basis.

Proof. Let $\boldsymbol{v}_1, \boldsymbol{v}_2, \ldots, \boldsymbol{v}_p$ be a spanning set for the vector space V, assumed not to be $\{\boldsymbol{0}\}$. If $\boldsymbol{v}_1, \boldsymbol{v}_2, \ldots, \boldsymbol{v}_p$ are linearly independent they form a basis. If not, by Theorem 2.1, one of the vectors $\boldsymbol{v}_k$ is a linear combination of the preceding ones. We eliminate this vector from the set and, renumbering the vectors if necessary, we obtain a set of $p - 1$ vectors $\boldsymbol{v}_1, \boldsymbol{v}_2, \ldots, \boldsymbol{v}_{p-1}$. Clearly any linear combination of $\boldsymbol{v}_1, \boldsymbol{v}_2, \ldots, \boldsymbol{v}_p$ is also a linear combination of $\boldsymbol{v}_1, \boldsymbol{v}_2, \ldots, \boldsymbol{v}_{p-1}$, so that the latter set is also a spanning set for V. Continuing in this way we arrive at a linearly independent spanning set $\boldsymbol{v}_1, \boldsymbol{v}_2, \ldots, \boldsymbol{v}_n$

$(1 \leqslant n \leqslant p)$ and this is a basis. Thus every spanning set of V contains a basis.

A vector space is *finite-dimensional* if it has a finite basis (i.e., a basis consisting of a finite number of vectors). It follows that any vector space that has a finite spanning set is finite-dimensional.

THEOREM 2.3. If V is a finite-dimensional vector space with a basis $\boldsymbol{v}_1, \boldsymbol{v}_2, \ldots, \boldsymbol{v}_n$ then every vector $\boldsymbol{v} \in V$ can be expressed in one and only one way as a linear combination $x_1\boldsymbol{v}_1 + x_2\boldsymbol{v}_2 + \cdots + x_n\boldsymbol{v}_n$.

Proof. Since a basis is a spanning set, $\boldsymbol{v}$ can certainly be expressed in at least one way as a linear combination of $\boldsymbol{v}_1, \boldsymbol{v}_2, \ldots, \boldsymbol{v}_n$. Suppose that

$$\boldsymbol{v} = x_1\boldsymbol{v}_1 + x_2\boldsymbol{v}_2 + \cdots + x_n\boldsymbol{v}_n = y_1\boldsymbol{v}_1 + y_2\boldsymbol{v}_2 + \cdots + y_n\boldsymbol{v}_n.$$

Then $$(x_1 - y_1)\boldsymbol{v}_1 + (x_2 - y_2)\boldsymbol{v}_2 + \cdots + (x_n - y_n)\boldsymbol{v}_n = \boldsymbol{0}.$$

But the vectors $\boldsymbol{v}_1, \boldsymbol{v}_2, \ldots, \boldsymbol{v}_n$ are linearly independent, so that this implies

$$x_1 - y_1 = x_2 - y_2 = \cdots = x_n - y_n = 0,$$

i.e., $$x_1 = y_1, x_2 = y_2, \ldots, x_n = y_n.$$

Thus every vector $\boldsymbol{v} \in V$ has a *unique* representation as a linear combination of a set of basis vectors.

THEOREM 2.4. All bases of a finite-dimensional vector space have the same number of elements.

Proof. Let $\{\boldsymbol{v}_1, \boldsymbol{v}_2, \ldots, \boldsymbol{v}_n\}$ and $\{\boldsymbol{w}_1, \boldsymbol{w}_2, \ldots, \boldsymbol{w}_m\}$ be two bases of the vector space V. $\boldsymbol{w}_m$ is a linear combination of the vectors $\boldsymbol{v}_1, \boldsymbol{v}_2, \ldots, \boldsymbol{v}_n$ so that the set $\{\boldsymbol{w}_m, \boldsymbol{v}_1, \boldsymbol{v}_2, \ldots, \boldsymbol{v}_n\}$ spans V and is linearly dependent. It follows from Theorem 2.1 that some vector $\boldsymbol{v}_i$ is a linear combination of the preceding ones. By renumbering the vectors $\boldsymbol{v}$ if necessary we can suppose $\boldsymbol{v}_n$ to be a linear combination of $\boldsymbol{w}_m, \boldsymbol{v}_1, \boldsymbol{v}_2, \ldots, \boldsymbol{v}_{n-1}$. It then follows that $\{\boldsymbol{w}_m, \boldsymbol{v}_1, \boldsymbol{v}_2, \ldots, \boldsymbol{v}_{n-1}\}$ is a spanning set for V. Again, $\{\boldsymbol{w}_{m-1}, \boldsymbol{w}_m, \boldsymbol{v}_1, \ldots, \boldsymbol{v}_{n-1}\}$ is a linearly dependent spanning set for V and, as before, we can eliminate one of the vectors $\boldsymbol{v}$, leaving a spanning set for V. With suitable renumbering we can eliminate $\boldsymbol{v}_{n-1}$, leaving a spanning set $\{\boldsymbol{w}_{m-1}, \boldsymbol{w}_m, \boldsymbol{v}_1, \ldots, \boldsymbol{v}_{n-2}\}$. Suppose $n < m$. We then arrive eventually at a spanning set $\{\boldsymbol{w}_{m-n+1}, \boldsymbol{w}_{m-n+2}, \ldots, \boldsymbol{w}_m\}$. But this is a contradiction, since $\boldsymbol{w}_1$, for example, cannot be expressed as a linear combination of $\boldsymbol{w}_{m-n+1}, \ldots,$

$\boldsymbol{w}_m$, the set of vectors $\boldsymbol{w}$ being linearly independent. It follows that n $\not< m$, i.e., $n \geqslant m$. Repeating the argument used above, but with the roles of $\boldsymbol{v}$ and $\boldsymbol{w}$ interchanged, we see that $m \geqslant n$. Hence $m = n$.

We can now define the dimension of a finite-dimensional vector space.

Definition. The *dimension* of a finite-dimensional vector space V, written dim V, is the number of elements in a basis.

THEOREM 2.5. Any set of linearly independent vectors in a finite-dimensional vector space is part of a basis of the space.

Proof. Let V be a vector space of dimension n, and let the vectors $\boldsymbol{a}_1, \boldsymbol{a}_2, \ldots, \boldsymbol{a}_r$ in V be linearly independent. Let $\boldsymbol{v}_1, \boldsymbol{v}_2, \ldots, \boldsymbol{v}_n$ be a basis of V. Then the set $\{\boldsymbol{a}_1, \boldsymbol{a}_2, \ldots, \boldsymbol{a}_r, \boldsymbol{v}_1, \boldsymbol{v}_2, \ldots, \boldsymbol{v}_n\}$ is a spanning set for V and, as in Theorem 2.2, we delete from this set in order any vector that is a linear combination of the preceding ones. In this way we arrive at a basis (which must necessarily contain n elements). Moreover, since $\boldsymbol{a}_1, \boldsymbol{a}_2, \ldots, \boldsymbol{a}_r$ are linearly independent, none of these is deleted and the resulting basis contains all of them.

We say that the set $\boldsymbol{a}_1, \boldsymbol{a}_2, \ldots, \boldsymbol{a}_r$ has been *extended* to form a basis of V.

Corollary 1. In an n-dimensional vector space any set of $n + 1$ vectors is linearly dependent.

If they are independent they are part of a basis and this basis must contain at least $n + 1$ elements, which is a contradiction.

Corollary 2. If M is a subspace of V, dim $M \leqslant$ dim V. Moreover if dim $M =$ dim V, then $M = V$.

Any basis of M is a linearly independent set in V and can be extended to form a basis of V. Therefore the number of elements in a basis of V is not less than the number in a basis of M. If dim $M =$ dim V, then every basis of M is also a basis of V, so that $M = V$.

2.6 Sum of two subspaces

Let M and N be two subspaces of a vector space V and define $M + N$ to be the set of all vectors $\boldsymbol{m} + \boldsymbol{n}$, where $\boldsymbol{m} \in M$ and $\boldsymbol{n} \in N$. Clearly $M + N$ is contained in V (or is possibly identical with V). We write this $M + N \subseteq V$.

Let $\boldsymbol{v}_1, \boldsymbol{v}_2 \in M + N$. Then we can write

$${}_1 = \boldsymbol{m}_1 + \boldsymbol{n}_1, \qquad \boldsymbol{v}_2 = \boldsymbol{m}_2 + \boldsymbol{n}_2,$$

where $$\boldsymbol{m}_1, \boldsymbol{m}_2 \in M, \qquad \boldsymbol{n}_1, \boldsymbol{n}_2 \in N.$$

Hence $$\boldsymbol{v}_1 + \boldsymbol{v}_2 = (\boldsymbol{m}_1 + \boldsymbol{n}_1) + (\boldsymbol{m}_2 + \boldsymbol{n}_2)$$
$$= (\boldsymbol{m}_1 + \boldsymbol{m}_2) + (\boldsymbol{n}_1 + \boldsymbol{n}_2).$$

But since M and N are subspaces, $\boldsymbol{m}_1 + \boldsymbol{m}_2 \in M$ and $\boldsymbol{n}_1 + \boldsymbol{n}_2 \in N$. Hence $\boldsymbol{v}_1 + \boldsymbol{v}_2 \in M + N$. Also, for any scalar x, $x\boldsymbol{v}_1 = x\boldsymbol{m}_1 + x\boldsymbol{n}_1 \in M + N$, since $x\boldsymbol{m}_1 \in M$ and $x\boldsymbol{n}_1 \in N$.

We have therefore shown that $M + N$ is a subspace of V. Since $\boldsymbol{0} \in N$, $\boldsymbol{m} = \boldsymbol{m} + \boldsymbol{0} \in M + N$ for every $\boldsymbol{m} \in M$, so that $M \subseteq M + N$ and similarly $N \subseteq M + N$.

Let T be any subspace of V containing both M and N. Then for every $\boldsymbol{m} \in M$ and $\boldsymbol{n} \in N$ we have $\boldsymbol{m} \in T$ and $\boldsymbol{n} \in T$, so that $\boldsymbol{m} + \boldsymbol{n} \in T$ and hence $M + N \subseteq T$. We have shown that every subspace containing both M and N also contains $M + N$, so that $M + N$ is the smallest subspace containing both M and N.

It is important to note that $M + N$ does not consist merely of the vectors in M together with those in N. We illustrate this by referring again to the 'ordinary' space of three-dimensional vectors $\boldsymbol{i}$, $\boldsymbol{j}$, $\boldsymbol{k}$. $[\boldsymbol{i}]$ is a one-dimensional subspace consisting of a certain line and $[\boldsymbol{j}]$ is another line, but $[\boldsymbol{i}, \boldsymbol{j}]$, the set of all vectors of the form $x\boldsymbol{i} + y\boldsymbol{j}$, is a two-dimensional subspace. It is the plane of $\boldsymbol{i}$ and $\boldsymbol{j}$, and not merely the two lines.

We have already shown that $M \cap N$ is a subspace of V. Now every subspace contains the zero vector $\boldsymbol{0}$ so that $M \cap N$ always has at least one element. If, however, M and N have only the vector $\boldsymbol{0}$ in common, so that $M \cap N = \{\boldsymbol{0}\}$, we say that the sum $M + N$ is a *direct sum* and we write it $M \oplus N$. We now establish a relationship between the dimensions of the subspaces M, N, $M \cap N$, $M + N$.

THEOREM 2.6. If M and N are subspaces of a finite-dimensional vector space V over the field F, then

$$\dim M + \dim N = \dim (M \cap N) + \dim (M + N).$$

Proof. Suppose that $M \cap N \neq \{\boldsymbol{0}\}$, and that $M \cap N$ has a basis $\{\boldsymbol{u}_1, \boldsymbol{u}_2, \ldots, \boldsymbol{u}_r\}$. Since $M \cap N \subseteq M$, this set of vectors forms part of a basis of M. Let $\{\boldsymbol{u}_1, \ldots, \boldsymbol{u}_r, \boldsymbol{v}_1, \ldots, \boldsymbol{v}_s\}$ be a basis of M. Similarly let $\{\boldsymbol{u}_1, \ldots, \boldsymbol{u}_r, \boldsymbol{w}_1, \ldots, \boldsymbol{w}_t\}$ be a basis of N. Thus $M \cap N$, M, N have dimensions r, $r + s$, $r + t$ respectively. Clearly $\{\boldsymbol{u}_1, \ldots, \boldsymbol{u}_r, \boldsymbol{v}_1,$

$\ldots, \boldsymbol{v}_s, \boldsymbol{w}_1, \ldots, \boldsymbol{w}_t\}$ is a spanning set for $M + N$, so that dim $(M + N) \leqslant r + s + t$. We now show that the equality must hold. This will follow if we can show that this spanning set is linearly independent and hence forms a basis for $M + N$. Suppose that

$$x_1\boldsymbol{u}_1 + \cdots + x_r\boldsymbol{u}_r + y_1\boldsymbol{v}_1 + \cdots + y_s\boldsymbol{v}_s + z_1\boldsymbol{w}_1 + \cdots + z_t\boldsymbol{w}_t = \boldsymbol{0} \tag{1}$$

where the x's, y's, and z's are all in F. Then

$$x_1\boldsymbol{u}_1 + \cdots + x_r\boldsymbol{u}_r + y_1\boldsymbol{v}_1 + \cdots + y_s\boldsymbol{v}_s = -(z_1\boldsymbol{w}_1 + \cdots + z_t\boldsymbol{w}_t).$$

Now the vector on the right is in N (since the $\boldsymbol{w}$'s are basis vectors of N) and that on the left is in M. It follows that each is in $M \cap N$ and is therefore of the form $\alpha_1\boldsymbol{u}_1 + \cdots + \alpha_r\boldsymbol{u}_r$. Thus

$$\alpha_1\boldsymbol{u}_1 + \cdots + \alpha_r\boldsymbol{u}_r + z_1\boldsymbol{w}_1 + \cdots + z_t\boldsymbol{w}_t = \boldsymbol{0}.$$

But the vectors $\boldsymbol{u}_1, \ldots, \boldsymbol{u}_r, \boldsymbol{w}_1, \ldots, \boldsymbol{w}_t$ are linearly independent since, by hypothesis, they form a basis for N. Hence

$$\alpha_1 = \cdots = \alpha_r = z_1 = \cdots = z_t = 0.$$

Equation (1) then becomes

$$x_1\boldsymbol{u}_1 + \cdots + x_r\boldsymbol{u}_r + y_1\boldsymbol{v}_1 + \cdots + y_s\boldsymbol{v}_s = \boldsymbol{0}$$

and again $x_1 = \cdots = x_r = y_1 = \cdots = y_s = 0$, since $\boldsymbol{u}_1, \ldots, \boldsymbol{u}_r, \boldsymbol{v}_1, \ldots, \boldsymbol{v}_s$ are linearly independent. Consequently, Eqn. (1) implies that all the x's, y's, and z's are zero. This means that $\boldsymbol{u}_1, \ldots, \boldsymbol{u}_r, \boldsymbol{v}_1, \ldots, \boldsymbol{v}_s, \boldsymbol{w}_1, \ldots, \boldsymbol{w}_t$ are linearly independent and hence form a basis for $M + N$. Thus

$$\begin{aligned}\dim (M + N) &= r + s + t = (r + s) + (r + t) - r \\ &= \dim M + \dim N - \dim (M \cap N)\end{aligned}$$

The result is now proved.

If $M \cap N = \{\boldsymbol{0}\}$, let $\{\boldsymbol{v}_1, \ldots, \boldsymbol{v}_s\}$ be a basis for M and $\{\boldsymbol{w}_1, \ldots, \boldsymbol{w}_t\}$ be a basis for N. Then $\{\boldsymbol{v}_1, \ldots, \boldsymbol{v}_s, \boldsymbol{w}_1, \ldots, \boldsymbol{w}_t\}$ is a spanning set for $M + N = M \oplus N$. But if

$$y_1\boldsymbol{v}_1 + \cdots + y_s\boldsymbol{v}_s + z_1\boldsymbol{w}_1 + \cdots + z_t\boldsymbol{w}_t = \boldsymbol{0},$$

then $$y_1\boldsymbol{v}_1 + \cdots + y_s\boldsymbol{v}_s = -(z_1\boldsymbol{w}_1 + \cdots + z_t\boldsymbol{w}_t) = \boldsymbol{0},$$

since each vector is in $M \cap N$. This in turn implies that

$$y_1 = \cdots = y_s = z_1 = \cdots = z_t = 0.$$

Hence $\{\boldsymbol{v}_1, \ldots, \boldsymbol{v}_s, \boldsymbol{w}_1, \ldots, \boldsymbol{w}_t\}$ is a basis for $M \oplus N$ and

$$\dim (M \oplus N) = s + t = \dim M + \dim N.$$

This agrees with the general result provided that

$$\dim (M \cap N) = \dim \{\mathbf{0}\} = 0.$$

We therefore agree to define the dimension of the space $\{\mathbf{0}\}$ to be zero.

Every vector $\mathbf{v} \in M \oplus N$ has a unique representation

$$\mathbf{v} = y_1\mathbf{v}_1 + \cdots + y_s\mathbf{v}_s + z_1\mathbf{w}_1 + \cdots + z_t\mathbf{w}_t = \mathbf{m} + \mathbf{n},$$

where $$\mathbf{m} = y_1\mathbf{v}_1 + \cdots + y_s\mathbf{v}_s \in M$$

and $$\mathbf{n} = z_1\mathbf{w}_1 + \cdots + z_t\mathbf{w}_t \in N.$$

This is not true if $M + N$ is not a direct sum. We can certainly write $\mathbf{v} = \mathbf{m} + \mathbf{n}$, where $\mathbf{m} \in M$, $\mathbf{n} \in N$, but the decomposition in this case is not unique.

2.7 Coordinates

We now show how to introduce coordinates into a vector space. Let V be an n-dimensional vector space over the field F and let $\{\mathbf{v}_1, \mathbf{v}_2, \ldots, \mathbf{v}_n\}$ be an *ordered* basis for V. The term 'ordered' implies that the order of the vectors is specified and that the same set of vectors in a different order is regarded as a different basis. Any vector $\mathbf{v} \in V$ can be expressed uniquely in the form

$$\mathbf{v} = x_1\mathbf{v}_1 + x_2\mathbf{v}_2 + \cdots + x_n\mathbf{v}_n,$$

where $x_i \in F$ $(1 \leqslant i \leqslant n)$ and $(x_1, x_2, \ldots, x_n)$ are called the *coordinates* of $\mathbf{v}$ relative to the given basis. There is clearly a one–one correspondence between vectors $\mathbf{v} \in V$ and ordered n-tuples $(x_1, x_2, \ldots, x_n)$ with elements in F. The reader should verify that if $\mathbf{v}$, $\mathbf{w}$ have coordinates $(x_1, x_2, \ldots, x_n)$, $(y_1, y_2, \ldots, y_n)$ respectively and $k \in F$, then $\mathbf{v} + \mathbf{w}$ has coordinates $(x_1 + y_1, x_2 + y_2, \ldots, x_n + y_n)$ and $k\mathbf{v}$ has coordinates $(kx_1, kx_2, \ldots, kx_n)$.

2.8 Further examples

In Section 2.2, example 1, we discussed the vector space V of all 2×2 matrices with real elements over the field $\mathbf{R}$. We see at once that the matrices

$$\begin{bmatrix} 1 & 0 \\ 0 & 0 \end{bmatrix}, \quad \begin{bmatrix} 0 & 1 \\ 0 & 0 \end{bmatrix}, \quad \begin{bmatrix} 0 & 0 \\ 1 & 0 \end{bmatrix}, \quad \begin{bmatrix} 0 & 0 \\ 0 & 1 \end{bmatrix}$$

form a spanning set for V. For any element $\begin{bmatrix} a & b \\ c & d \end{bmatrix} \in V$ can be written

$$\begin{bmatrix} a & b \\ c & d \end{bmatrix} = a\begin{bmatrix} 1 & 0 \\ 0 & 0 \end{bmatrix} + b\begin{bmatrix} 0 & 1 \\ 0 & 0 \end{bmatrix} + c\begin{bmatrix} 0 & 0 \\ 1 & 0 \end{bmatrix} + d\begin{bmatrix} 0 & 0 \\ 0 & 1 \end{bmatrix}.$$

Moreover, these four matrices are linearly independent. For

$$\begin{bmatrix} a & b \\ c & d \end{bmatrix} = \begin{bmatrix} 0 & 0 \\ 0 & 0 \end{bmatrix}$$

if and only if $a = b = c = d = 0$. It follows that dim $V = 4$ and the four given matrices form a basis for the space.

In Section 2.2, example 3, we discussed the vector space F_n. The vectors $\boldsymbol{e}_1 = (1, 0, 0, \ldots, 0)$, $\boldsymbol{e}_2 = (0, 1, 0, \ldots, 0)$, $\ldots$, $\boldsymbol{e}_n = (0, 0, 0, \ldots, 0, 1)$ form a basis for the space, since

$$(x_1, x_2, \ldots, x_n) = x_1\boldsymbol{e}_1 + x_2\boldsymbol{e}_2 + \cdots + x_n\boldsymbol{e}_n$$

and the vector on the right is zero if and only if $x_1 = x_2 = \cdots = x_n = 0$. This is called the *standard basis* for F_n. Consider the space $\boldsymbol{Q}_3$ of triples with rational coordinates. $\boldsymbol{e}_1 = (1, 0, 0)$, $\boldsymbol{e}_2 = (0, 1, 0)$, $\boldsymbol{e}_3 = (0, 0, 1)$ is the standard basis for the space, which has dimension 3. On the other hand, $\boldsymbol{v}_1 = (1, 0, 2)$, $\boldsymbol{v}_2 = (-1, 1, 0)$, $\boldsymbol{v}_3 = (0, 2, 3)$ is also a basis for the space. For, if

$$x(1, 0, 2) + y(-1, 1, 0) + z(0, 2, 3) = (0, 0, 0),$$

we have $\quad x - y = 0, \qquad y + 2z = 0, \qquad 2x + 3z = 0,$

giving $\quad x = y = z = 0,$

so that $\boldsymbol{v}_1$, $\boldsymbol{v}_2$, $\boldsymbol{v}_3$ are linearly independent.

$$\boldsymbol{u} = (1, -1, 1) = 2\boldsymbol{v}_1 + \boldsymbol{v}_2 - \boldsymbol{v}_3$$

and $\quad \boldsymbol{w} = (-1, 8, 11) = \boldsymbol{v}_1 + 2\boldsymbol{v}_2 + 3\boldsymbol{v}_3.$

Hence the vectors $\boldsymbol{u}$, $\boldsymbol{w}$ have coordinates $(2, 1, -1)$, $(1, 2, 3)$ respectively relative to the basis $\{\boldsymbol{v}_1, \boldsymbol{v}_2, \boldsymbol{v}_3\}$. Note that the basis vectors $\boldsymbol{v}_1$, $\boldsymbol{v}_2$, $\boldsymbol{v}_3$ have coordinates $(1, 0, 0)$, $(0, 1, 0)$, $(0, 0, 1)$ respectively.

As a final example, consider solutions of the ordinary differential equation

$$d^2y/dx^2 - 4dy/dx - 5y = 0.$$

If $y_1(x)$ and $y_2(x)$ are solutions of the equation, then $y_1(x) + y_2(x)$ is also a solution and $ky_1(x)$ is a solution for any complex number k. It is easy to see that the associative laws are satisfied and that solutions of the equation form a vector space (called the solution space). It can be shown that the solution space of an ordinary differential equation of order n, having constant coefficients, is of dimension n. Therefore, if we can, in any way whatsoever, find n linearly independent solutions $y_1, y_2, \ldots, y_n$, then every solution of the equation can be expressed in the form

$$y = c_1y_1 + c_2y_2 + \cdots + c_ny_n,$$

where $c_1, c_2, \ldots, c_n$ are scalars (i.e., complex numbers). In the given example e^{5x} and e^{-x} are linearly independent solutions and the general solution is

$$y = ae^{5x} + be^{-x},$$

where a, b are arbitrary scalars.

Problems

2.1 Show that the set of complex numbers $\boldsymbol{C}$ is a vector space over the real numbers $\boldsymbol{R}$. Show that 1, i (where $i^2 = -1$) form a basis for the space.

2.2 Prove that the set of all numbers of the form $a + b\sqrt{2} + c\sqrt{5}$, where $a, b, c \in \boldsymbol{Q}$, is a vector space V over $\boldsymbol{Q}$, and find a basis for the space. Find a subspace S of dimension 2 and find another subspace T such that $S \oplus T = V$.

2.3 Show that the vectors $\boldsymbol{v}_1 = (1, 2, 0, 3)$, $\boldsymbol{v}_2 = (2, -1, 0, 0)$, span a two-dimensional subspace of $\boldsymbol{Q}_4$ and extend $\boldsymbol{v}_1, \boldsymbol{v}_2$ to form a basis of $\boldsymbol{Q}_4$ by adjoining two of the standard basis vectors.

2.4 Prove that the set of all 2×2 matrices with complex elements is a vector space over $\boldsymbol{R}$. Find a basis for the space and hence determine its dimension.

2.5 Prove that the vectors $(1, 2, 0)$, $(0, 5, 7)$, $(-1, 1, 3)$ form a basis for $\boldsymbol{R}_3$ and find the coordinates of the vectors $(0, 13, 17)$ and $(2, 3, 1)$ relative to this basis.

2.6 Prove that a field F may be regarded as a vector space over itself and find its dimension.

2.7 Determine k so that the vectors $(1, -1, k - 1)$, $(2, k, -4)$, $(0, k + 2, -8)$ in $\boldsymbol{R}_3$ are linearly dependent.

2.8 $\boldsymbol{x} = (x_1, x_2, x_3, x_4)$ is an element in the space $\boldsymbol{Q}_4$. M is the set of all elements for which $x_1 = x_2$ and N is the set of all elements for which $x_4 = 0$. Prove that M and N are subspaces of $\boldsymbol{Q}_4$ and find the dimension of each. Find $\dim (M \cap N)$ and $\dim (M + N)$ and verify the dimension formula of Theorem 2.6.

2.9 Deduce from the axioms for a vector space V over the field F that (i) $0\boldsymbol{v} = \boldsymbol{o}$, (ii) $(-1)\boldsymbol{v} = -\boldsymbol{v}$, (iii) $x \,.\, \boldsymbol{o} = \boldsymbol{o}$, where $\boldsymbol{v} \in V$, $x \in F$.

2.10 $C[0, 1]$ is the set of all functions that are continuous on the interval $[0, 1]$. Show that, with a suitable definition of addition and scalar multiplication, $C[0, 1]$ is a real vector space. Let M be the subset of $C[0, 1]$ consisting of all functions f for which $f(0) = f(1)$. Prove that M is a subspace of $C[0, 1]$.

2.11 Prove that the set of all polynomials in X of degree not exceeding n, with real coefficients, forms a real vector space (with the usual definition of addition and scalar multiplication). Find a basis for the space and hence determine its dimension.

2.12 V is the real vector space formed by 2×2 matrices with real elements (see Section 2.2, example 1). U is the subset of V consisting of all matrices of the form $\begin{bmatrix} a & 0 \\ 0 & b \end{bmatrix}$ and W is the subset consisting of all matrices of the form $\begin{bmatrix} c & 0 \\ d & 0 \end{bmatrix}$. Prove that U and W are subspaces of V and find a basis for each. Find a basis for $U \cap W$ and one for $U + W$ and hence verify the dimension theorem.

2.13 The subspace U of $\boldsymbol{R}_4$ is spanned by the vectors $(1, 0, 2, 3)$ and $(0, 1, -1, 2)$ and the subspace V is spanned by $(1, 2, 3, 4)$, $(-1, -1, 5, 0)$, and $(0, 0, 0, 1)$. Find the dimensions of U, V, $U \cap V$ and $U + V$.

2.14 Prove that the solutions of the equations

$$\begin{aligned} 2x_1 - 3x_2 + 5x_3 &= 0 \\ x_1 + 4x_2 - x_3 &= 0 \end{aligned}$$

form a subspace of $\boldsymbol{R}_3$.

3 · *Linear Transformations*

Applications of vector spaces are usually concerned with mappings which associate with a vector in one space a unique vector either in the same space or in another space. The mappings, or transformations, which interest us are called linear. Before giving a general definition we shall consider an example.

Let Ox, Oy, Oz be rectangular cartesian axes in a three-dimensional space and let (x, y, z) be the coordinates of the point P referred to these axes. The point P has a unique projection P′ on the plane of Ox, Oy and P′ has coordinates (x, y) referred to these axes. Thus we may regard $(x, y, z) \rightarrow (x, y)$ as a mapping from the space $\boldsymbol{R}_3$ to the space $\boldsymbol{R}_2$.

If P_1, P_2 have coordinates (x_1, y_1, z_1), (x_2, y_2, z_2) respectively, then the projections P_1', P_2' have coordinates (x_1, y_1), (x_2, y_2) respectively. Clearly the projection of the point $(x_1 + x_2, y_1 + y_2, z_1 + z_2)$ is the point $(x_1 + x_2, y_1 + y_2)$ so that, with an obvious notation,

$$(x_1, y_1, z_1) + (x_2, y_2, z_2) \rightarrow (x_1, y_1) + (x_2, y_2).$$

Again, $c(x, y, z) \rightarrow c(x, y)$, where c is real. This is an example of a linear transformation and we shall now give a general definition.

3.1 Definition of a linear transformation

Let U and V be two vector spaces over the same field F and let T be a mapping that assigns to every $\boldsymbol{u} \in U$ a unique vector $T(\boldsymbol{u}) \in V$ such that

$$T(\boldsymbol{u}_1 + \boldsymbol{u}_2) = T(\boldsymbol{u}_1) + T(\boldsymbol{u}_2) \quad \text{for all} \quad \boldsymbol{u}_1, \boldsymbol{u}_2 \in U$$

and $$T(c\boldsymbol{u}) = cT(\boldsymbol{u}) \quad \text{for all} \quad \boldsymbol{u} \in U \quad \text{and all} \quad c \in F.$$

Then T is called a *linear transformation* (or a *linear mapping*) of U into V. We denote by $L(U, V)$ the set of all linear mappings of U into V, so that $T \in L(U, V)$. If $T(\boldsymbol{u}) = \boldsymbol{v}$ we call $\boldsymbol{v}$ the *image* of $\boldsymbol{u}$ under T. Note that if U and V are different spaces they need not have the same dimension. In the example mentioned above U, V have dimension 3, 2 respectively. If $U = V$, i.e., $T \in L(V, V)$, we say that T is a linear transformation on V. T is said to be *one–one* if distinct (i.e., different) elements in U have distinct images in V, and T is said to be *onto* if every element in V is the image of an element in U. In this case the image space is the whole of V. If T is both one–one and onto it is said to be *non-singular*. This means that for each $\boldsymbol{v} \in V$ there exists a *unique* $\boldsymbol{u} \in U$ such that $T(\boldsymbol{u}) = \boldsymbol{v}$, and we write $\boldsymbol{u} = T^{-1}(\boldsymbol{v})$. Let

$$T(\boldsymbol{u}_1) = \boldsymbol{v}_1, \qquad T(\boldsymbol{u}_2) = \boldsymbol{v}_2.$$

Then $$\boldsymbol{v}_1 + \boldsymbol{v}_2 = T(\boldsymbol{u}_1) + T(\boldsymbol{u}_2) = T(\boldsymbol{u}_1 + \boldsymbol{u}_2)$$

and $$c\boldsymbol{v}_1 = cT(\boldsymbol{u}_1) = T(c\boldsymbol{u}_1).$$

Thus $$\boldsymbol{u}_1 + \boldsymbol{u}_2 = T^{-1}(\boldsymbol{v}_1) + T^{-1}(\boldsymbol{v}_2) = T^{-1}(\boldsymbol{v}_1 + \boldsymbol{v}_2)$$

and $$c\boldsymbol{u}_1 = cT^{-1}(\boldsymbol{v}_1) = T^{-1}(c\boldsymbol{v}_1).$$

It follows that T^{-1} is a linear mapping in $L(V, U)$ and we call it the *inverse* of T. If we define the mapping $T^{-1}T$ so that $(T^{-1}T)(\boldsymbol{u}) = T^{-1}\{T(\boldsymbol{u})\}$ for each $\boldsymbol{u} \in U$, it is easily seen that $T^{-1}T$ is the identity mapping on the space U (i.e., the mapping that leaves every vector in U unchanged). Similarly TT^{-1} is the identity mapping on V.

3.2 Matrix of a linear transformation

To simplify our calculations, we can introduce coordinates. Let U, V be vector spaces of dimension m, n respectively over a field F and let $T \in L(U, V)$. We choose an ordered basis $\{\boldsymbol{a}_1, \boldsymbol{a}_2, \ldots, \boldsymbol{a}_m\}$ for U and an ordered basis $\{\boldsymbol{b}_1, \boldsymbol{b}_2, \ldots, \boldsymbol{b}_n\}$ for V. Now $T(\boldsymbol{a}_i) \in V$ for $i = 1, 2, \ldots, m$ and every element of V can be expressed uniquely as a linear combination of the basis vectors $\boldsymbol{b}_1, \boldsymbol{b}_2, \ldots, \boldsymbol{b}_n$. Hence

$$T(\boldsymbol{a}_i) = \sum_{j=1}^{n} t_{ji}\boldsymbol{b}_j \qquad (i = 1, 2, \ldots, m),$$

where the scalars t_{ji} $(i = 1, 2, \ldots, m;\ j = 1, 2, \ldots, n)$ all belong to the field F. The set of elements t_{ji} determines a matrix $\boldsymbol{T}$ of order $n \times m$ given by

$$\boldsymbol{T} = \begin{bmatrix} t_{11} & t_{12} \cdot\cdot\cdot & t_{1m} \\ t_{21} & t_{22} \cdot\cdot\cdot & t_{2m} \\ \cdot & \cdot\ \cdot\ \cdot & \cdot \\ t_{n1} & t_{n2} \cdot\cdot\cdot & t_{nm} \end{bmatrix}.$$

The elements of this matrix depend on the two bases taken for U and V, and for any given mapping T they are uniquely determined by the bases. The matrix $\boldsymbol{T}$ is called the *matrix of the transformation T relative to the given bases*. Let us now find the image of a given vector $\boldsymbol{u} \in U$ under the mapping T. Now $\boldsymbol{u}$ can be written uniquely as

$$\boldsymbol{u} = u_1\boldsymbol{a}_1 + u_2\boldsymbol{a}_2 + \cdots + u_m\boldsymbol{a}_m,$$

so that $\boldsymbol{u}$ has coordinate vector $(u_1, u_2, \ldots, u_m)$ relative to the given basis. Hence

$$\begin{aligned} T(\boldsymbol{u}) &= u_1T(\boldsymbol{a}_1) + u_2T(\boldsymbol{a}_2) + \cdots + u_mT(\boldsymbol{a}_m) \\ &= u_1\sum_{j=1}^{n} t_{j1}\boldsymbol{b}_j + u_2\sum_{j=1}^{n} t_{j2}\boldsymbol{b}_j + \cdots + u_m\sum_{j=1}^{n} t_{jm}\boldsymbol{b}_j \\ &= \sum_{k=1}^{m} u_k \sum_{j=1}^{n} t_{jk}\boldsymbol{b}_j \\ &= \sum_{j=1}^{n}\left(\sum_{k=1}^{m} t_{jk}u_k\right)\boldsymbol{b}_j. \end{aligned}$$

But $T(\boldsymbol{u}) \in V$, so that $T(\boldsymbol{u})$ has a unique coordinate vector $(v_1, v_2, \ldots, v_n)$ relative to the basis $\{\boldsymbol{b}_1, \boldsymbol{b}_2, \ldots, \boldsymbol{b}_n\}$. From the above equation we see that

$$v_j = \sum_{k=1}^{m} t_{jk}u_k \qquad (j = 1, 2, \ldots, n). \tag{1}$$

We shall return to this equation and put it into the form of a matrix equation.

3.3 Operations on matrices

Here is a reminder of the rules for adding and multiplying matrices (for further details see, for example, Tropper, Ref. 12). If $\boldsymbol{A}$ and $\boldsymbol{B}$ are two matrices of the *same order* $m \times n$, their sum $\boldsymbol{A} + \boldsymbol{B}$ exists and is also an $m \times n$ matrix. Each element of $\boldsymbol{A} + \boldsymbol{B}$ is equal to the sum of the corresponding elements of $\boldsymbol{A}$ and $\boldsymbol{B}$. Thus

$$\boldsymbol{A} + \boldsymbol{B} = \begin{bmatrix} a_{11} + b_{11} & a_{12} + b_{12} & \cdots & a_{1n} + b_{1n} \\ a_{21} + b_{21} & a_{22} + b_{22} & \cdots & a_{2n} + b_{2n} \\ \cdot & \cdot & \cdots & \cdot \\ a_{m1} + b_{m1} & a_{m2} + b_{m2} & \cdots & a_{mn} + b_{mn} \end{bmatrix}.$$

The transpose of $\boldsymbol{A}$, denoted by $\boldsymbol{A}^{\mathrm{t}}$, is the matrix obtained from $\boldsymbol{A}$ by interchanging its rows and its columns. Thus $\boldsymbol{A}^{\mathrm{t}}$ is of order $n \times m$ and

$$\boldsymbol{A}^{\mathrm{t}} = \begin{bmatrix} a_{11} & a_{21} \cdots a_{m1} \\ a_{12} & a_{22} \cdots a_{m2} \\ \cdot & \cdot \ \cdot \ \cdot \ \cdot \\ a_{1n} & a_{2n} \cdots a_{mn} \end{bmatrix}.$$

The ijth element (i.e., the element in the ith row and the jth column) of $\boldsymbol{A}^{\mathrm{t}}$ is the jith element of $\boldsymbol{A}$ and we write $[\boldsymbol{A}^{\mathrm{t}}]_{ij} = a_{ji}$. If $\boldsymbol{A}$ and $\boldsymbol{B}$ are of *different* orders their sum does not exist. If k is a scalar, then $k\boldsymbol{A}$ is the matrix obtained from $\boldsymbol{A}$ by multiplying every element by k. Thus

$$k\boldsymbol{A} = \begin{bmatrix} ka_{11} & ka_{12} \cdots ka_{1n} \\ ka_{21} & ka_{22} \cdots ka_{2n} \\ \cdot & \cdot \ \cdot \ \cdot \ \cdot \\ ka_{m1} & ka_{m2} \cdots ka_{mn} \end{bmatrix}.$$

If $\boldsymbol{A}$ is of order $m \times n$ and $\boldsymbol{B}$ is of order $n \times p$, the product $\boldsymbol{AB}$ is of order $m \times p$. The ijth element of $\boldsymbol{AB}$, written $[\boldsymbol{AB}]_{ij}$, is given by

$$[\boldsymbol{AB}]_{ij} = \sum_{j=1}^{n} a_{ik}b_{kj}. \tag{2}$$

As a special case, suppose that $\boldsymbol{A}$ is of order $1 \times n$, so that it consists of a single row $[a_1, a_2, \ldots, a_n]$. It is then called a *row vector*. Let $\boldsymbol{B}$ be of order $n \times 1$, so that it consists of a single column $\begin{bmatrix} b_1 \\ b_2 \\ \cdot \\ \cdot \\ b_n \end{bmatrix}$. $\boldsymbol{B}$ is then called a *column vector*, which is usually written $\{b_1, b_2, \ldots, b_n\}$ for convenience in printing. Then $\boldsymbol{AB}$ is of order 1×1 and consists of a single element. Thus

$$\boldsymbol{AB} = a_1b_1 + a_2b_2 + \cdots + a_nb_n. \tag{3}$$

The expression on the right is called the *inner product* of the row vector $[a_1, a_2, \ldots, a_n]$ and the column vector $\{b_1, b_2, \ldots, b_n\}$. If we denote column vectors by single letters in boldface type, so that

$$\boldsymbol{b} = \{b_1\, b_2 \ldots b_n\}, \qquad \boldsymbol{a} = \{a_1\, a_2 \ldots a_n\}$$

then, since a row vector is the transpose of a column vector,

$$\boldsymbol{a}^{\mathrm{t}} = [a_1 \quad a_2 \cdots a_n].$$

Equation (3) can now be written

$$\boldsymbol{a}^{t}\boldsymbol{b} = a_1b_1 + a_2b_2 + \cdots + a_nb_n = \boldsymbol{b}^{t}\boldsymbol{a}.$$

The inner product of two vectors exists *only* if they are of the same order, i.e., if they contain the same number of elements. We see now that Eqn. (2) implies

$$[\boldsymbol{AB}]_{ij} = \text{inner product of } i\text{th row of } \boldsymbol{A} \text{ and } j\text{th column of } \boldsymbol{B}.$$

The product does not exist unless the number of columns in $\boldsymbol{A}$ is equal to the number of rows in $\boldsymbol{B}$.

In particular, if $\boldsymbol{A}$ is a matrix of order $n \times m$ and $\boldsymbol{u} = \{u_1\, u_2 \cdots u_m\}$ is a column vector of order m (i.e., a matrix of order $m \times 1$), then $\boldsymbol{Au}$ is a column vector $\boldsymbol{v} = \{v_1\, v_2 \cdots v_n\}$ of order n, where

$$\begin{aligned} v_j &= \text{inner product of } j\text{th row vector of } \boldsymbol{A} \text{ and column vector } \boldsymbol{u} \\ &= \sum_{k=1}^{m} a_{jk}u_k. \end{aligned}$$

We now return to Eqn. (1) of Section 3.2, namely

$$v_j = \sum_{k=1}^{m} t_{jk}u_k \qquad (j = 1, 2, \ldots, n).$$

Writing $\boldsymbol{u} = \{u_1\, u_2 \cdots u_m\}$, $\boldsymbol{v} = \{v_1\, v_2 \cdots v_n\}$ and

$$\boldsymbol{T} = \begin{bmatrix} t_{11} & t_{12} \cdots t_{1m} \\ t_{21} & t_{22} \cdots t_{2m} \\ \cdot & \cdot\ \cdot\ \cdot\ \cdot\ \cdot \\ t_{n1} & t_{n2} \cdots t_{nm} \end{bmatrix}.$$

we see that this set of equations is equivalent to the single matrix equation

$$\boldsymbol{v} = \boldsymbol{Tu}.$$

3.4 Product of two transformations

Let U, V, W be vector spaces of dimension m, n, p respectively and let T, S be two linear transformations, where $T \in L(U, V)$, $S \in L(V, W)$. Thus for every $\boldsymbol{u} \in U$ there exists a unique $\boldsymbol{v} \in V$ such that $T(\boldsymbol{u}) = \boldsymbol{v}$ and there exists a unique $\boldsymbol{w} \in W$ such that $S(\boldsymbol{v}) = \boldsymbol{w}$.

If we combine the two mappings, we see that each $\boldsymbol{u} \in U$ corresponds to a unique $\boldsymbol{w} \in W$. We call the composite transformation ST, so that

$$(ST)(\boldsymbol{u}) = S\{T(\boldsymbol{u})\}.$$

If $T(\boldsymbol{u}_1) = \boldsymbol{v}_1$, $T(\boldsymbol{u}_2) = \boldsymbol{v}_2$, $S(\boldsymbol{v}_1) = \boldsymbol{w}_1$, $S(\boldsymbol{v}_2) = \boldsymbol{w}_2$, then

$$T(\boldsymbol{u}_1 + \boldsymbol{u}_2) = \boldsymbol{v}_1 + \boldsymbol{v}_2 \qquad \text{and} \qquad S(\boldsymbol{v}_1 + \boldsymbol{v}_2) = \boldsymbol{w}_1 + \boldsymbol{w}_2,$$

so that $$ST(\boldsymbol{u}_1 + \boldsymbol{u}_2) = \boldsymbol{w}_1 + \boldsymbol{w}_2 = ST(\boldsymbol{u}_1) + ST(\boldsymbol{u}_2).$$

We also have, for any scalar c,

$$ST(c\boldsymbol{u}) = S\{cT(\boldsymbol{u})\} = S(c\boldsymbol{v}) = cS(\boldsymbol{v}) = c\boldsymbol{w} = cST(\boldsymbol{u}).$$

Thus ST is a linear transformation belonging to $L(U, W)$. We regard this as being the *product* of the two transformations S and T and note that TS does not exist if $W \neq U$.

We now choose ordered bases $\{\boldsymbol{a}_1, \boldsymbol{a}_2, \ldots, \boldsymbol{a}_m\}$, $\{\boldsymbol{b}_1, \boldsymbol{b}_2, \ldots, \boldsymbol{b}_n\}$, $\{\boldsymbol{c}_1, \boldsymbol{c}_2, \ldots, \boldsymbol{c}_p\}$ for U, V, W respectively. As in Section 3.3 we can determine the matrices $\boldsymbol{T}$, $\boldsymbol{S}$ of the mapping T, S respectively relative to these bases, T is of order $n \times m$, S is of order $p \times n$. Now

$$T(\boldsymbol{a}_i) = \sum_{j=1}^{n} t_{ji}\boldsymbol{b}_j \qquad (i = 1, 2, \ldots, m)$$

and $$S(\boldsymbol{b}_j) = \sum_{k=1}^{k} s_{kj}\boldsymbol{c}_k \qquad (j = 1, 2, \ldots, n)$$

so that
$$\begin{aligned}(ST)(\boldsymbol{a}_i) &= S\left(\sum_{j=1}^{n} t_{ji}\boldsymbol{b}_j\right)\\ &= \sum_{j=1}^{n} t_{ji}S(\boldsymbol{b}_j)\\ &= \sum_{j=1}^{n} t_{ji}\sum_{k=1}^{k} s_{kj}\boldsymbol{c}_k\\ &= \sum_{j=1}^{p}\left(\sum_{k=1}^{n} s_{kj}t_{ji}\right)\boldsymbol{c}_k \qquad (i = 1, 2, \ldots, m).\end{aligned}$$

The matrix of the product transformation ST therefore has as its kith element $\sum_{j=1}^{n} s_{kj}t_{ji}$. But this is the kith element of the product matrix $\boldsymbol{ST}$, so that the transformation ST has matrix $\boldsymbol{ST}$.

3.5 Inverse transformations

Let U and V have the same dimension n and let T be a linear transformation in $L(U, V)$ and S be a linear transformation in $L(V, U)$. Then

$ST \in L(U, U)$. Now one transformation in $L(U, U)$ is the identity transformation 1, which maps each element $\boldsymbol{u} \in U$ onto itself. If, for a given $T \in L(U, V)$, there exists an inverse transformation $T^{-1} \in L(V, U)$ such that $T^{-1}T = 1$, we say that T is *non-singular*. Relative to given bases $\{\boldsymbol{a}_1, \boldsymbol{a}_2, \ldots, \boldsymbol{a}_n\}$ of U and $\{\boldsymbol{b}_1, \boldsymbol{b}_2, \ldots, \boldsymbol{b}_n\}$ of V, T and T^{-1} both have square matrices $\boldsymbol{T}$ and $\boldsymbol{T}^{-1}$ of order $n \times n$. The identity transformation 1 in $L(U, U)$ is such that

$$1(\boldsymbol{a}_i) = \boldsymbol{a}_i \qquad (i = 1, 2, \ldots, n)$$

so that 1 has the $n \times n$ matrix $\boldsymbol{I}_n$ given by

$$\boldsymbol{I}_n = \begin{bmatrix} 1 & 0 \cdots 0 \\ 0 & 1 \cdots 0 \\ \cdot & \cdots\cdots \\ 0 & 0 \cdots 1 \end{bmatrix}.$$

Using the Kronecker delta defined by

$$\delta_{ij} = \begin{cases} 1 & (i = j) \\ 0 & (i \neq j) \end{cases}$$

we have
$$[\boldsymbol{I}_n]_{ij} = \delta_{ij}.$$

The matrix $\boldsymbol{I}_n$ is called the unit matrix, since for every $m \times n$ matrix $\boldsymbol{A}$ and every $n \times p$ matrix $\boldsymbol{B}$,

$$\boldsymbol{A}\boldsymbol{I}_n = \boldsymbol{A}, \qquad \boldsymbol{I}_n\boldsymbol{B} = \boldsymbol{B}.$$

There are unit matrices $\boldsymbol{I}_1, \boldsymbol{I}_2, \ldots, \boldsymbol{I}_n, \ldots$ of all orders, but when there is no ambiguity we omit the subscript and write $\boldsymbol{I}$. Now the matrix of the transformation $T^{-1}T$ is $\boldsymbol{T}^{-1}\boldsymbol{T} = \boldsymbol{I}$. But a matrix $\boldsymbol{S}$ such that $\boldsymbol{ST} = \boldsymbol{I}$ is called the *inverse* (or reciprocal) of $\boldsymbol{T}$ and we write $\boldsymbol{S} = \boldsymbol{T}^{-1}$. Thus if a transformation is non-singular, the matrix of the inverse transformation is the inverse of the matrix of the transformation. A square matrix $\boldsymbol{T}$ is said to be *non-singular* if it has an inverse $\boldsymbol{T}^{-1}$. It is easy to see that if $\boldsymbol{T}^{-1}\boldsymbol{T} = \boldsymbol{I}$, then $\boldsymbol{T}\boldsymbol{T}^{-1} = \boldsymbol{I}$ also, and that the inverse $\boldsymbol{T}^{-1}$ is unique. First, note that multiplication of matrices is associative. If $\boldsymbol{A}, \boldsymbol{B}, \boldsymbol{C}$ are of order $m \times n, n \times p, p \times q$ respectively, then $(\boldsymbol{AB})\boldsymbol{C} = \boldsymbol{A}(\boldsymbol{BC})$ and we can write the product $\boldsymbol{ABC}$ without ambiguity.

Suppose that $\boldsymbol{T}^{-1}\boldsymbol{T} = \boldsymbol{I}$ and that $\boldsymbol{TS} = \boldsymbol{I}$. Then

$$(\boldsymbol{T}^{-1}\boldsymbol{T})\boldsymbol{S} = \boldsymbol{IS} = \boldsymbol{S}.$$

But
$$\boldsymbol{T}^{-1}(\boldsymbol{TS}) = \boldsymbol{T}^{-1}\boldsymbol{I} = \boldsymbol{T}^{-1},$$

so that
$$\boldsymbol{S} = \boldsymbol{T}^{-1} \quad \text{and} \quad \boldsymbol{T}\boldsymbol{T}^{-1} = \boldsymbol{I}.$$

If $\boldsymbol{PT} = \boldsymbol{I}$, then $\boldsymbol{PT} - \boldsymbol{T}^{-1}\boldsymbol{T} = \boldsymbol{0}$, the zero matrix on the right being the $n \times n$ matrix whose elements are all equal to zero. Hence

$$(\boldsymbol{P} - \boldsymbol{T}^{-1})\boldsymbol{T} = \boldsymbol{0},$$
$$(\boldsymbol{P} - \boldsymbol{T}^{-1})\boldsymbol{T}\boldsymbol{T}^{-1} = \boldsymbol{0}\boldsymbol{T}^{-1} = \boldsymbol{0},$$
$$\boldsymbol{P} - \boldsymbol{T}^{-1} = \boldsymbol{0},$$
$$\boldsymbol{P} = \boldsymbol{T}^{-1}.$$

Thus $\boldsymbol{T}^{-1}\boldsymbol{T} = \boldsymbol{T}\boldsymbol{T}^{-1} = \boldsymbol{I}$ and $\boldsymbol{T}^{-1}$ is unique.

3.6 Change of basis

Let U be a vector space of dimension m and let $\{\boldsymbol{a}_1, \boldsymbol{a}_2, \ldots, \boldsymbol{a}_m\}$, $\{\boldsymbol{c}_1, \boldsymbol{c}_2, \ldots, \boldsymbol{c}_m\}$ be two bases of U. Each vector $\boldsymbol{c}_i$ is a unique linear combination of the vectors $\boldsymbol{a}_1, \boldsymbol{a}_2, \ldots, \boldsymbol{a}_m$, so that

$$\boldsymbol{c}_i = \sum_{j=1}^{m} p_{ji}\boldsymbol{a}_j \qquad (i = 1, 2, \ldots, m).$$

If we regard $[\boldsymbol{c}_1\ \boldsymbol{c}_2 \cdots \boldsymbol{c}_m]$ as a row vector whose elements are themselves vectors, we can write these equations in matrix form

$$[\boldsymbol{c}_1 \quad \boldsymbol{c}_2 \cdots \boldsymbol{c}_m] = [\boldsymbol{a}_1 \quad \boldsymbol{a}_2 \cdots \boldsymbol{a}_m]\boldsymbol{P},$$

where $\boldsymbol{P} = [p_{ji}]$.

We may regard $\boldsymbol{P}$ as the matrix of a linear transformation belonging to $L(U, U)$. But each $\boldsymbol{a}_i$ is a unique linear combination of the vectors $\boldsymbol{c}_1, \boldsymbol{c}_2, \ldots, \boldsymbol{c}_m$, so that the mapping is one–one and $\boldsymbol{P}$ has an inverse. The matrix $\boldsymbol{P}$ is thus non-singular. If the vector $\boldsymbol{u} \in U$ has coordinate vector $\boldsymbol{x} = (x_1, x_2, \ldots, x_m)$ with respect to the basis $\{\boldsymbol{a}_1, \boldsymbol{a}_2, \ldots, \boldsymbol{a}_m\}$ and coodinate vector $\boldsymbol{y} = (y_1, y_2, \ldots, y_m)$ with respect to the basis $\{\boldsymbol{c}_1, \boldsymbol{c}_2, \ldots, \boldsymbol{c}_m\}$, then

$$\boldsymbol{u} = \sum_{i=1}^{m} y_i\boldsymbol{c}_i = \sum_{i=1}^{m} y_i \sum_{j=1}^{m} p_{ji}\boldsymbol{a}_j$$
$$= \sum_{j=1}^{m}\left(\sum_{i=1}^{m} p_{ji}y_i\right)\boldsymbol{a}_j = \sum_{j=1}^{m} x_j\boldsymbol{a}_j.$$

Hence $$x_j = \sum_{i=1}^{m} p_{ji}y_i \qquad (j = 1, 2, \ldots, m)$$

and $$\boldsymbol{x} = \boldsymbol{P}\boldsymbol{y}.$$

Similarly, let V be a vector space of dimension n and let $\{\boldsymbol{b}_1, \boldsymbol{b}_2, \ldots, \boldsymbol{b}_n\}$, $\{\boldsymbol{d}_1, \boldsymbol{d}_2, \ldots, \boldsymbol{d}_n\}$ be two bases of V. Then there exists a non-singular matrix $\boldsymbol{Q}$ such that

$$\boldsymbol{d}_i = \sum_{j=1}^{n} q_{ji}\boldsymbol{b}_j.$$

This may also be written

$$\boldsymbol{b}_i = \sum_{j=1}^{n} q_{ji}^{-1}\boldsymbol{d}_j,$$

where $[\boldsymbol{Q}^{-1}]_{ji} = q_{ji}^{-1}$. Moreover, if the vector $\boldsymbol{v} \in V$ has coordinate vectors $\boldsymbol{z} = (z_1, z_2, \ldots, z_n)$, $\boldsymbol{w} = (w_1, w_2, \ldots, w_n)$ with respect to the bases $\{\boldsymbol{b}_1, \boldsymbol{b}_2, \ldots, \boldsymbol{b}_n\}$, $\{\boldsymbol{d}_1, \boldsymbol{d}_2, \ldots, \boldsymbol{d}_n\}$ respectively, then

$$\boldsymbol{z} = \boldsymbol{Q}\boldsymbol{w}.$$

Now let T be a transformation belonging to $L(U, V)$ and let it have matrix $\boldsymbol{T} = [t_{ji}]$ relative to the bases $\{\boldsymbol{a}_1, \boldsymbol{a}_2, \ldots, \boldsymbol{a}_m\}$, $\{\boldsymbol{b}_1, \boldsymbol{b}_2, \ldots, \boldsymbol{b}_n\}$ for U, V respectively. Thus

$$T(\boldsymbol{a}_i) = \sum_{j=1}^{n} t_{ji}\boldsymbol{b}_j \qquad (i = 1, 2, \ldots, m)$$

so that

$$\begin{aligned} T(\boldsymbol{c}_k) &= T\left(\sum_{i=1}^{m} p_{ik}\boldsymbol{a}_i\right) = \sum_{i=1}^{m} p_{ik}T(\boldsymbol{a}_i) \\ &= \sum_{i=1}^{m} p_{ik}\sum_{j=1}^{n} t_{ji}\boldsymbol{b}_j = \sum_{j=1}^{n}\left(\sum_{i=1}^{m} t_{ji}p_{ik}\right)\boldsymbol{b}_j \\ &= \sum_{j=1}^{n} [\boldsymbol{TP}]_{jk}\sum_{l=1}^{n} q_{lj}^{-1}\boldsymbol{d}_l = \sum_{l=1}^{n} [\boldsymbol{Q}^{-1}\boldsymbol{TP}]_{lk}\boldsymbol{d}_l. \end{aligned}$$

Writing $\boldsymbol{Q}^{-1}\boldsymbol{TP} = \boldsymbol{S}$, we have

$$T(\boldsymbol{c}_k) = \sum_{l=1}^{n} s_{lk}\boldsymbol{d}_l,$$

and $\boldsymbol{S}$ is the matrix of the transformation T relative to the bases $\{\boldsymbol{c}_1, \boldsymbol{c}_2, \ldots, \boldsymbol{c}_m\}$, $\{\boldsymbol{d}_1, \boldsymbol{d}_2, \ldots, \boldsymbol{d}_n\}$ for U, V respectively. $\boldsymbol{Q}$ is non-singular, so that $\boldsymbol{Q}^{-1}$ (which has $\boldsymbol{Q}$ as its inverse) is also non-singular.

Consider the special case $U = V$, so that $T \in L(U, U)$, and let T have matrix $\boldsymbol{T}$ relative to the basis $\{\boldsymbol{a}_1, \boldsymbol{a}_2, \ldots, \boldsymbol{a}_m\}$ for U. Let $\boldsymbol{S}$ be the matrix of T relative to the basis $\{\boldsymbol{c}_1, \boldsymbol{c}_2, \ldots, \boldsymbol{c}_m\}$ for U where, as before,

$$[\boldsymbol{c}_1 \quad \boldsymbol{c}_2 \cdots \boldsymbol{c}_m] = [\boldsymbol{a}_1 \quad \boldsymbol{a}_2 \cdots \boldsymbol{a}_m]\boldsymbol{P},$$

$\boldsymbol{P}$ being a non-singular $m \times m$ matrix. In this case $\boldsymbol{Q} = \boldsymbol{P}$ and therefore

$$\boldsymbol{S} = \boldsymbol{P}^{-1}\boldsymbol{T}\boldsymbol{P}.$$

This result can be restated as a theorem as follows.

THEOREM 3.1. If $T \in L(U, V)$ has matrices $\boldsymbol{T}$, $\boldsymbol{S}$ with respect to two pairs of bases of U, V, then there exist non-singular matrices $\boldsymbol{P}$, $\boldsymbol{Q}$ such that $\boldsymbol{S} = \boldsymbol{Q}^{-1}\boldsymbol{T}\boldsymbol{P}$. In particular, if $U = V$, then $\boldsymbol{Q} = \boldsymbol{P}$.

3.7 Determinant of a linear transformation

We define the determinant of a square matrix $\boldsymbol{A}$, denoted by det $\boldsymbol{A}$ or $|\boldsymbol{A}|$, to be the determinant having exactly the same rows and columns as $\boldsymbol{A}$. Since the rule (or one of the rules) for multiplying determinants is the same as that for matrices, the determinant of the product of two square matrices of the same order is the product of their determinants. In particular, if $\boldsymbol{A}^{-1}$ exists,

$$\det \boldsymbol{A} \cdot \det \boldsymbol{A}^{-1} = \det \boldsymbol{A}\boldsymbol{A}^{-1} = \det \boldsymbol{I} = 1,$$

so that $\det \boldsymbol{A} \neq 0$ and $\det \boldsymbol{A}^{-1} = 1/\det \boldsymbol{A}$.

Let $T \in L(V, V)$ and let T have matrices $\boldsymbol{T}$, $\boldsymbol{S}$ with respect to two different bases of V. Then, by Theorem 3.1, $\boldsymbol{S} = \boldsymbol{P}^{-1}\boldsymbol{T}\boldsymbol{P}$, where $\boldsymbol{P}$ is non-singular. Hence

$$\begin{aligned} \det \boldsymbol{S} &= \det \boldsymbol{P}^{-1}\boldsymbol{T}\boldsymbol{P} = \det \boldsymbol{P}^{-1} \cdot \det \boldsymbol{T} \cdot \det \boldsymbol{P} \\ &= \det \boldsymbol{T}, \end{aligned}$$

since $\det \boldsymbol{P}^{-1} = 1/\det \boldsymbol{P}$.

It follows that the matrix of a given linear transformation relative to any basis always has the same determinant, and we can therefore define the *determinant of a linear transformation* to be the determinant of the matrix of the transformation relative to any basis.

Example 1. V is the vector space consisting of all polynomials in X with coefficients in the field F, having degree less than n. D is the differentiation operator d/dX. Since $D(f + g) = Df + Dg$ and $D(cf) = cD(f)$ for all polynomials f, g and all $c \in F$, D is a linear transformation on V, i.e., $D \in L(V, V)$.

Every polynomial in V can be written in the form

$$a_0 + a_1X + a_2X^2 + \cdots + a_{n-1}X^{n-1}$$

and we can therefore take $1, X, X^2, \ldots, X^{n-1}$ as a basis for V. Now

$$D(1) = 0,$$
$$D(X^k) = kX^{k-1} \qquad (k = 1, 2, \ldots, n-1).$$

Thus the matrix of the transformation relative to the above basis is the $n \times n$ matrix

$$A = \begin{bmatrix} 0 & 1 & 0 & 0 \cdots 0 & \\ 0 & 0 & 2 & 0 \cdots 0 & \\ 0 & 0 & 0 & 3 \cdots 0 & \\ \cdot & \cdot & \cdot & \cdots\cdots & \\ 0 & 0 & 0 & 0 \cdots n-1 & \\ 0 & 0 & 0 & 0 \cdots 0 & \end{bmatrix}.$$

Now if $\boldsymbol{B}$ is *any* $n \times n$ matrix, the first column of the product matrix $\boldsymbol{BA}$ consists entirely of zeros. There is therefore *no* matrix $\boldsymbol{B}$ such that $\boldsymbol{BA} = \boldsymbol{I}$ and it follows that $\boldsymbol{A}$ has no inverse and is singular. This is also clear from the fact that $D(f + c) = Df$ for all scalars c, so that the mapping is not one–one.

Consider the special case $n = 3$. Then V has a basis $\{1, X, X^2\}$ and

$$A = \begin{bmatrix} 0 & 1 & 0 \\ 0 & 0 & 2 \\ 0 & 0 & 0 \end{bmatrix}$$

relative to this basis. We now find the matrix of D relative to the new basis $\{1 + X, 1 - X, 1 + X + X^2\}$. These three elements are linearly independent, for if

$$a(1 + X) + b(1 - X) + c(1 + X + X^2) = 0 = 0 + 0X + 0X^2$$

i.e., $$cX^2 + (c + a - b)X + (c + a + b) = 0,$$

we have $\quad c = 0, \qquad c + a - b = c + a + b = 0, \qquad a = b = 0.$

They therefore form a basis for V. Now

$$[1 + X \quad 1 - X \quad 1 + X + X^2] = [1 \quad X \quad X^2] \begin{bmatrix} 1 & 1 & 1 \\ 1 & -1 & 1 \\ 0 & 0 & 1 \end{bmatrix}.$$

Thus $$P = \begin{bmatrix} 1 & 1 & 1 \\ 1 & -1 & 1 \\ 0 & 0 & 1 \end{bmatrix}, \qquad P^{-1} = \begin{bmatrix} \frac{1}{2} & \frac{1}{2} & -1 \\ \frac{1}{2} & -\frac{1}{2} & 0 \\ 0 & 0 & 1 \end{bmatrix}.$$

If $\boldsymbol{B}$ is the matrix of the mapping D relative to the new basis,

$$\begin{aligned}\boldsymbol{B} &= \boldsymbol{P}^{-1}\boldsymbol{A}\boldsymbol{P}\\ &= \begin{bmatrix} \frac{1}{2} & \frac{1}{2} & -1 \\ \frac{1}{2} & -\frac{1}{2} & 0 \\ 0 & 0 & 1 \end{bmatrix} \begin{bmatrix} 0 & 1 & 0 \\ 0 & 0 & 2 \\ 0 & 0 & 0 \end{bmatrix} \begin{bmatrix} 1 & 1 & 1 \\ 1 & -1 & 1 \\ 0 & 0 & 1 \end{bmatrix}\\ &= \begin{bmatrix} 0 & \frac{1}{2} & 1 \\ 0 & \frac{1}{2} & -1 \\ 0 & 0 & 0 \end{bmatrix} \begin{bmatrix} 1 & 1 & 1 \\ 1 & -1 & 1 \\ 0 & 0 & 1 \end{bmatrix}\\ &= \begin{bmatrix} \frac{1}{2} & -\frac{1}{2} & \frac{3}{2} \\ \frac{1}{2} & -\frac{1}{2} & -\frac{1}{2} \\ 0 & 0 & 0 \end{bmatrix}.\end{aligned}$$

We can check this result as follows. Write

$$c_1 = 1 + X, \qquad c_2 = 1 - X, \qquad c_3 = 1 + X + X^2.$$

$$\begin{aligned}Dc_1 &= 1 = \tfrac{1}{2}(c_1 + c_2)\\ &= [c_1 \quad c_2 \quad c_3] \begin{bmatrix} \frac{1}{2} \\ \frac{1}{2} \\ 0 \end{bmatrix},\end{aligned}$$

$$Dc_2 = -1 = -\tfrac{1}{2}(c_1 + c_2) = [c_1 \quad c_2 \quad c_3] \begin{bmatrix} -\frac{1}{2} \\ -\frac{1}{2} \\ 0 \end{bmatrix}$$

$$Dc_3 = 2X + 1 = \tfrac{2}{3}c_1 - \tfrac{1}{2}c_2 = [c_1 \quad c_2 \quad c_3] \begin{bmatrix} \frac{3}{2} \\ -\frac{1}{2} \\ 0 \end{bmatrix},$$

and combining these we obtain

$$[Dc_1 \quad Dc_2 \quad Dc_3] = [c_1 \quad c_2 \quad c_3]\boldsymbol{B}.$$

Example 2. Let T be the mapping from $\boldsymbol{R}_3$ into $\boldsymbol{R}_2$ given by $T(v_1, v_2, v_3) = (v_1, v_2)$. Now

$$\begin{aligned}T[(a_1, a_2, a_3) + (b_1, b_2, b_3)] &= T(a_1 + b_1, a_2 + b_2, a_3 + b_3)\\ &= (a_1 + b_1, a_2 + b_2)\\ &= (a_1, a_2) + (b_1, b_2)\\ &= T(a_1, a_2, a_3) + T(b_1, b_2, b_3)\end{aligned}$$

and

$$\begin{aligned}T[c(a_1, a_2, a_3)] &= T(ca_1, ca_2, ca_3)\\ &= (ca_1, ca_2)\\ &= cT(a_1, a_2, a_3).\end{aligned}$$

Hence T is a linear mapping, i.e., $T \in L(\boldsymbol{R}_3, \boldsymbol{R}_2)$. This mapping is singular, since $T(v_1, v_2, a) = (v_1, v_2)$ for all $a \in \boldsymbol{R}$ and the inverse does not exist. We can give the mapping a geometrical interpretation. The point (x, y) is the projection of the point (x, y, z) on the x, y-plane, the coordinates being rectangular cartesian. $\boldsymbol{R}_3$ has a basis $\{\boldsymbol{e}_1, \boldsymbol{e}_2, \boldsymbol{e}_3\}$, where $\boldsymbol{e}_1 = (1, 0, 0)$, $\boldsymbol{e}_2 = (0, 1, 0)$, $\boldsymbol{e}_3 = (0, 0, 1)$. $\boldsymbol{R}_2$ has a basis $\{\boldsymbol{f}_1, \boldsymbol{f}_2\}$, where $\boldsymbol{f}_1 = (1, 0)$, $\boldsymbol{f}_2 = (0, 1)$. Clearly

$$\begin{aligned} T(\boldsymbol{e}_1) &= \boldsymbol{f}_1 = 1\boldsymbol{f}_1 + 0\boldsymbol{f}_2, \\ T(\boldsymbol{e}_2) &= \boldsymbol{f}_2 = 0\boldsymbol{f}_1 + 1\boldsymbol{f}_2, \\ T(\boldsymbol{e}_3) &= \boldsymbol{0} = 0\boldsymbol{f}_1 + 0\boldsymbol{f}_2 \end{aligned}$$

and the matrix of T is given by

$$\boldsymbol{T} = \begin{bmatrix} 1 & 0 & 0 \\ 0 & 1 & 0 \end{bmatrix}.$$

Problems

3.1 The set $\boldsymbol{C}$ of complex numbers is a vector space V over the field of real numbers $\boldsymbol{R}$. T is a mapping such that, for each $z \in \boldsymbol{C}$, $T(z)$ is the complex conjugate of z. Prove that T is a linear transformation on V, and find its matrix relative to the basis $\{1, i\}$ for V $(i^2 = -1)$. Show that the mapping is non-singular and find its inverse.

3.2 V is an n-dimensional vector space with basis $\{\boldsymbol{v}_1, \boldsymbol{v}_2, \ldots, \boldsymbol{v}_n\}$ and M is the p-dimensional subspace of V spanned by $\{\boldsymbol{v}_1, \boldsymbol{v}_2, \ldots, \boldsymbol{v}_p\}$, where $p < n$. T is a linear transformation on V that maps every element in M onto an element in M. Prove that the matrix of T can be partitioned in the form

$$\boldsymbol{T} = \begin{bmatrix} \boldsymbol{T}_1 & \boldsymbol{T}_3 \\ \boldsymbol{0} & \boldsymbol{T}_2 \end{bmatrix}$$

and find the order of the zero matrix $\boldsymbol{0}$.

3.3 The matrix of a linear transformation T on $\boldsymbol{R}_3$ relative to the basis $\{\boldsymbol{e}_1, \boldsymbol{e}_2, \boldsymbol{e}_3\}$ is

$$\begin{bmatrix} 0 & 1 & 1 \\ 1 & 0 & -1 \\ -1 & -1 & 0 \end{bmatrix}.$$

Find the matrix of T relative to the basis $\{(0, 1, 2), (1, 1, 1), (1, 0, -2)\}$.

3.4 A linear transformation T on $\boldsymbol{R}_2$ takes the vector $(2, 6)$ into $(1, 7)$ and the vector $(-1, 2)$ into $(4, 1)$. Find the matrix of this transformation relative to the standard basis.

3.5 P is the set of all polynomials p in x with coefficients in the field F. Prove that P is a vector space over F.

Mappings S and T on P are defined by $S(p) = p'$ (the derivative of p), $T(p) = xp$. Show that S and T are linear and determine the mapping $ST - TS$.

3.6 Determine which of the following transformations T are linear.

(i) $T\colon \boldsymbol{R}_3 \to \boldsymbol{R}_2$ defined by $T(x_1, x_2, x_3) = (x_2, x_3)$

(ii) $T\colon \boldsymbol{R}_3 \to \boldsymbol{R}_3$ defined by $T(x_1, x_2, x_3) = (x_1, x_2, x_3) + (1, 1, 1)$

(iii) $T\colon \boldsymbol{R}_2 \to \boldsymbol{R}_2$ defined by $T(x_1, x_2) = (x_1 - x_2, x_1 + x_2)$

(iv) $T\colon \boldsymbol{R}_n \to \boldsymbol{R}_n$ defined by $T\boldsymbol{x} = -\boldsymbol{x}$

(v) $T\colon \boldsymbol{R}_2 \to \boldsymbol{R}_2$ defined by $T(x_1, x_2) = (x_2, x_1)$

(vi) $T\colon \boldsymbol{R}_2 \to \boldsymbol{R}$ defined by $T(x_1, x_2) = x_1x_2$.

3.7 Solve the equations

$$\begin{aligned} x_1 + 2x_2 - 2x_3 &= y_1 \\ x_1 + 5x_2 + 3x_3 &= y_2 \\ 2x_1 + 6x_2 - x_3 &= y_3 \end{aligned}$$

giving x_1, x_2, x_3 in terms of y_1, y_2, y_3. Hence write down the inverse of the matrix

$$A = \begin{bmatrix} 1 & 2 & -2 \\ 1 & 5 & 3 \\ 2 & 6 & -1 \end{bmatrix}$$

and verify that it is the inverse.

3.8 P_n is the real vector space of polynomials p with real coefficients, of degree not exceeding n. T is a mapping of P_n into itself given by

$$T\{p(x)\} = p(x + 1).$$

Prove that T is linear.

3.9 V is the set of all real functions that are continuous everywhere. Prove that V is a real vector space and prove that W, the set of all real functions having continuous second derivatives everywhere, is a subspace of V.

T is a mapping from W into V given by

$$T\{f(x)\} = af''(x) + bf'(x) + cf(x),$$

where a, b, c are real constants. Prove that T is linear.

3.10 A and B, both in $L(\boldsymbol{R}_3, \boldsymbol{R}_2)$, are given by

$$\begin{aligned} A(x_1, x_2, x_3) &= (x_1 + x_2 - x_3, 2x_2 + x_3), \\ B(x_1, x_2, x_3) &= (x_2, x_3). \end{aligned}$$

Write down

(*a*) $A(2, -5, 7)$

(*b*) $B(4, 6, -3)$

(*c*) $(A + B)(x_1, x_2, x_3)$

(*d*) $(A - 2B)(1, 2, -1)$

Write down the matrices of A and B relative to the standard bases.

3.11 A and B, both in $L(\boldsymbol{R}_3, \boldsymbol{R}_3)$, are given by

$$\begin{aligned} A(x_1, x_2, x_3) &= (x_2 + x_3, x_3 + x_1, x_1 + x_2), \\ B(x_1, x_2, x_3) &= (x_1 - x_2, x_2 - x_3, x_3 - x_1). \end{aligned}$$

Calculate $(AB)(x_1, x_2, x_3)$ and $(BA)(x_1, x_2, x_3)$.

4 · *Rank and Nullity*

In pure mathematics and in the application of mathematics to physical problems one of the most commonly occurring types of equation is the homogeneous linear equation. This is of the form

$$T(\boldsymbol{u}) = \boldsymbol{0},$$

where $\boldsymbol{u}$ is an element of the vector space U and T is a linear transformation on U. An ordinary differential equation

$$a_0 y^{(n)} + a_1 y^{(n-1)} + \cdots + a_n y = 0$$

with constant coefficients $a_0, a_1, \ldots, a_n$ is of this form and we have already seen that the solutions of such an equation form a vector space. A set of homogeneous linear equations can be written in matrix form

$$\boldsymbol{Tu} = \boldsymbol{0}.$$

For any given linear transformation T, the set of all vectors $T(\boldsymbol{u})$ ($\boldsymbol{u} \in U$) and the set of vectors of U for which $T(\boldsymbol{u}) = \boldsymbol{0}$ are of such importance that we study them in this chapter in some detail. We first show that both sets are vector spaces.

4.1 Kernel and image spaces

Let $T \in L(U, V)$ be a given linear transformation. The set of all vectors $\boldsymbol{u} \in U$ such that $T(\boldsymbol{u}) = \boldsymbol{0}$ is called the *kernel* of T and written ker T. The set of all vectors $T(\boldsymbol{u})$, $\boldsymbol{u} \in U$, is called the *image* of U under T and written $T(U)$.

THEOREM 4.1. Ker T is a subspace of U and $T(U)$ is a subspace of V, where $T \in L(U, V)$.

Proof. Let $\boldsymbol{u}_1, \boldsymbol{u}_2 \in$ ket T and $c \in F$. Then

$$T(\boldsymbol{u}_1) = \boldsymbol{0}, \qquad T(\boldsymbol{u}_2) = \boldsymbol{0}.$$

Hence $$T(\boldsymbol{u}_1 + \boldsymbol{u}_2) = T(\boldsymbol{u}_1) + T(\boldsymbol{u}_2) = \boldsymbol{0}$$

and $$T(c\boldsymbol{u}_1) = cT(\boldsymbol{u}_1) = \boldsymbol{0},$$

so that $\boldsymbol{u}_1 + \boldsymbol{u}_2 \in \ker T$ and $c\boldsymbol{u}_1 \in \ker T$. Ker T is therefore a subspace of U.

Again, if $\boldsymbol{v}_1, \boldsymbol{v}_2 \in T(U)$, there exist $\boldsymbol{u}_1, \boldsymbol{u}_2 \in U$ such that $\boldsymbol{v}_1 = T(\boldsymbol{u}_1)$, $\boldsymbol{v}_2 = T(\boldsymbol{u}_2)$. Then

$$\boldsymbol{v}_1 + \boldsymbol{v}_2 = T(\boldsymbol{u}_1) + T(\boldsymbol{u}_2) = T(\boldsymbol{u}_1 + \boldsymbol{u}_2)$$

and $$c\boldsymbol{v}_1 = cT(\boldsymbol{u}_1) = T(c\boldsymbol{u}_1).$$

But $\boldsymbol{u}_1 + \boldsymbol{u}_2$ and $c\boldsymbol{u}_1$ both belong to U. Hence $\boldsymbol{v}_1 + \boldsymbol{v}_2$ and $c\boldsymbol{v}_1$ both belong to $T(U)$. $T(U)$ is therefore a subspace of V.

THEOREM 4.2. dim ker T + dim $T(U)$ = dim U.

Proof. Let dim $U = n$, dim ker $T = s$, so that $s \leqslant n$. Let $\{\boldsymbol{u}_1, \boldsymbol{u}_2, \ldots, \boldsymbol{u}_s\}$ be a basis for ker T and extend this to a basis $\{\boldsymbol{u}_1, \ldots, \boldsymbol{u}_s, \boldsymbol{u}_{s+1}, \ldots, \boldsymbol{u}_n\}$ of U. We now show that $T(\boldsymbol{u}_{s+1}), \ldots, T(\boldsymbol{u}_n)$ is a basis of $T(U)$.

For any vector of $T(U)$ is of the form $T(\boldsymbol{u})$, $\boldsymbol{u} \in U$, and any vector $\boldsymbol{u}$ is of the form

$$\boldsymbol{u} = c_1\boldsymbol{u}_1 + c_2\boldsymbol{u}_2 + \cdots + c_n\boldsymbol{u}_n, \qquad (c_1, c_2, \ldots, c_n \in F).$$

Thus
$$\begin{aligned} T(\boldsymbol{u}) &= T(c_1\boldsymbol{u}_1) + \cdots + T(c_s\boldsymbol{u}_s) \\ &\qquad + T(c_{s+1}\boldsymbol{u}_{s+1}) + \cdots + T(c_n\boldsymbol{u}_n) \\ &= c_1T(\boldsymbol{u}_1) + \cdots + c_sT(\boldsymbol{u}_s) + c_{s+1}T(\boldsymbol{u}_{s+1}) \\ &\qquad + \cdots + c_nT(\boldsymbol{u}_n) \\ &= c_{s+1}T(\boldsymbol{u}_{s+1}) + \cdots + c_nT(\boldsymbol{u}_n), \end{aligned}$$

since $T(\boldsymbol{u}_1) = \cdots = T(\boldsymbol{u}_s) = \boldsymbol{0}$.

We have shown that any vector of $T(U)$ is a linear combination of the set $T(\boldsymbol{u}_{s+1}), \ldots, T(\boldsymbol{u}_n)$ which is therefore a spanning set for $T(U)$. But this spanning set is linearly independent. For if

$$c_{s+1}T(\boldsymbol{u}_{s+1}) + \cdots + c_nT(\boldsymbol{u}_n) = \boldsymbol{0},$$

then $$T(c_{s+1}\boldsymbol{u}_{s+1} + \cdots + c_n\boldsymbol{u}_n) = 0$$

and $$c_{s+1}\boldsymbol{u}_{s+1} + \cdots + c_n\boldsymbol{u}_n \in \ker T.$$

Thus $c_{s+1}\boldsymbol{u}_{s+1} + \cdots + c_n\boldsymbol{u}_n$ is a linear combination of $\boldsymbol{u}_1, \boldsymbol{u}_2, \ldots, \boldsymbol{u}_s$, giving a relation of the form

$$c_1\boldsymbol{u}_1 + \cdots + c_s\boldsymbol{u}_s + c_{s+1}\boldsymbol{u}_{s+1} + \cdots + c_n\boldsymbol{u}_n = \boldsymbol{0}.$$

Since $\boldsymbol{u}_1, \boldsymbol{u}_2, \ldots, \boldsymbol{u}_n$ are linearly independent, this implies that $c_1 = \cdots = c_s = c_{s+1} = \cdots = c_n = 0$. We have therefore shown that

$$c_{s+1}T(\boldsymbol{u}_{s+1}) + \cdots + c_nT(\boldsymbol{u}_n) = \boldsymbol{0}$$

only if $c_{s+1} = \cdots = c_n = 0$, and hence that $T(\boldsymbol{u}_{s+1}), \ldots, T(\boldsymbol{u}_n)$ is a linearly independent set. But it is a spanning set for $T(U)$ and is therefore a basis for $T(U)$, and $\dim T(U) = n - s$, i.e., $\dim T(U) = \dim U - \dim \ker T$, as required.

Definition. The dimension of ker T is called the *nullity* of T, and the dimension of $T(U)$ is called the *rank* of T. The above theorem can then be stated as follows.

If T is a linear transformation of a vector space U, then

$$\text{rank of } T + \text{nullity of } T = \dim U.$$

4.2 Normal forms

For a given linear transformation $T \in L(U, V)$ we naturally seek bases for U and V relative to which the matrix of T takes the simplest possible form. Such a form is called a *normal form*. Before proving Theorem 4.3 we must remember that, using the notation of partitioned matrices (see e.g., Tropper, Ref. 12), $\begin{bmatrix} \boldsymbol{I}_r & \boldsymbol{0} \\ \boldsymbol{0} & \boldsymbol{0} \end{bmatrix}$ denotes a matrix $\boldsymbol{A}$ in which all the elements are zero except for the r elements $a_{11}, a_{22}, \ldots, a_{rr}$ which are all equal to 1. The matrices

$$\begin{bmatrix} 1 & 0 & 0 & 0 \\ 0 & 1 & 0 & 0 \\ 0 & 0 & 0 & 0 \\ 0 & 0 & 0 & 0 \end{bmatrix}, \quad \begin{bmatrix} 1 & 0 & 0 & 0 & 0 & 0 \\ 0 & 1 & 0 & 0 & 0 & 0 \\ 0 & 0 & 1 & 0 & 0 & 0 \\ 0 & 0 & 0 & 0 & 0 & 0 \\ 0 & 0 & 0 & 0 & 0 & 0 \end{bmatrix}, \quad \begin{bmatrix} 1 & 0 & 0 & 0 \\ 0 & 0 & 0 & 0 \\ 0 & 0 & 0 & 0 \end{bmatrix}$$

are all of this form. Note that, although the unit matrix $\boldsymbol{I}_r$ is always square, the three zero submatrices are not necessarily square.

THEOREM 4.3. Given any linear transformation $T \in L(U, V)$, of rank r, we can always choose bases for U and V relative to which the matrix of T has the form $\begin{bmatrix} I_r & 0 \\ 0 & 0 \end{bmatrix}$.

Proof. Let U, V have dimension m, n respectively. By Theorems 4.1 and 4.2, ker T is a subspace of U and

$$\dim \ker T = m - r.$$

Hence we can choose a basis $\{a_{r+1}, \ldots, a_m\}$ for ker T and extend it to form a basis $\{a_1, \ldots, a_r, a_{r+1}, \ldots, a_m\}$ for U. Now

$$T(a_k) = 0 \qquad (k = r + 1, \ldots, m). \tag{1}$$

Write

$$T(a_k) = b_k \qquad (k = 1, \ldots, r). \tag{2}$$

As in Theorem 4.2 we can show that $\{b_1, \ldots, b_r\}$ is a basis for the image space $T(U)$.

We now find the matrix for T relative to the bases $\{a_1, \ldots, a_m\}$ for U and $\{b_1, \ldots, b_r\}$ for V. Equations (1) and (2) show that it has the required form $\begin{bmatrix} I_r & 0 \\ 0 & 0 \end{bmatrix}$.

Corollary. If A is the matrix of a linear transformation of rank r, then there exist non-singular matrices H, K such that

$$H^{-1}AK = \begin{bmatrix} I_r & 0 \\ 0 & 0 \end{bmatrix}.$$

This follows at once from the theorem, using Theorem 3.1.

4.3 Rank of a matrix

The maximum number of linearly independent rows of a matrix A is called the *row rank* of A and the maximum number of linearly independent columns is the *column rank* of A. We first show that these two ranks are equal.

THEOREM 4.4. The row and column ranks of a matrix are equal.

Proof. Let A be a matrix of order $n \times m$ and let its column rank be r. Its columns are all vectors of order n. Let them be $a_1, a_2, \ldots, a_m$. Among these are r linearly independent columns. Let us select such a set and label them in order $\boldsymbol{a}_{k_1}, \boldsymbol{a}_{k_2}, \ldots, \boldsymbol{a}_{k_r}$. Then every column can be expressed as a linear combination of these r columns, so that

$$\boldsymbol{a}_i = \sum_{h=1}^{r} p_{hi}\boldsymbol{a}_{k_h} \qquad (i = 1, 2, \ldots, m).$$

Each side of the above equation is a vector, and equating the jth elements of the equal vectors we obtain

$$a_{ji} = \sum_{h=1}^{r} p_{hi}a_{j,k_h} \qquad (i = 1, \ldots, m;\ j = 1, \ldots, n). \tag{3}$$

Now, the rows of $\boldsymbol{A}$ are all vectors of order m. Denote them by $\boldsymbol{a}^1, \boldsymbol{a}^2, \ldots, \boldsymbol{a}^n$. Then a_{ji} is the ith element of the row vector $\boldsymbol{a}^j$, and p_{hi} is the ith element of the row vector $[p_{h1}, p_{h2}, \ldots, p_{hr}] = \boldsymbol{p}^h$, say. We may therefore rewrite Eqn. (3) in the form

$$\boldsymbol{a}^j = \sum_{h=1}^{r} \boldsymbol{p}^h a_{j,k_h} \qquad (j = 1, \ldots, n).$$

Thus each row $\boldsymbol{a}^j$ of $\boldsymbol{A}$ can be expressed as a linear combination of the r vectors $\boldsymbol{p}^1, \boldsymbol{p}^2, \ldots, \boldsymbol{p}^r$, and it follows that the subspace spanned by the rows of $\boldsymbol{A}$ is of dimension $\leqslant r$. The number of linearly independent rows does not, therefore, exceed the number of linearly independent columns. Consequently for any matrix $\boldsymbol{A}$

$$\text{row rank of } \boldsymbol{A} \leqslant \text{column rank of } \boldsymbol{A}. \tag{4}$$

But we can apply the same argument to $\boldsymbol{A}^{\mathrm{t}}$;

$$\text{row rank of } \boldsymbol{A}^{\mathrm{t}} \leqslant \text{column rank of } \boldsymbol{A}^{\mathrm{t}}. \tag{5}$$

Now

$$\text{row rank of } \boldsymbol{A}^{\mathrm{t}} = \text{column rank of } \boldsymbol{A}$$

and

$$\text{column rank of } \boldsymbol{A}^{\mathrm{t}} = \text{row rank of } \boldsymbol{A},$$

so that (5) becomes

$$\text{column rank of } \boldsymbol{A} \leqslant \text{row rank of } \boldsymbol{A}.$$

This, together with (4), gives

$$\text{column rank of } \boldsymbol{A} = \text{row rank of } \boldsymbol{A}, \text{ as required.}$$

We can therefore define the *rank* of a matrix to be equal to the common value of the row rank and the column rank, and we write it $r(\boldsymbol{A})$. If $\boldsymbol{A}$ is of order $n \times m$, clearly $r \leqslant n$, $r \leqslant m$ and $r(\boldsymbol{A}) = r(\boldsymbol{A}^{\mathrm{t}})$. We now show the connection between the rank of a linear transformation and the rank of its matrix.

THEOREM 4.5. Let $T \in L(U, V)$ be a linear transformation and let $\boldsymbol{T}$ be its matrix relative to some pair of bases. Then the rank of T is equal to the rank of $\boldsymbol{T}$.

Proof. Let $u \in U$, $T(u) = v \in V$ have coordinate vectors $\boldsymbol{u} = (u_1, u_2, \ldots, u_m)$, $\boldsymbol{v} = (v_1, v_2, \ldots, v_n)$ respectively, relative to the given bases. Then, as in Section 3.3,

$$\boldsymbol{v} = \boldsymbol{T}\boldsymbol{u}.$$

Writing $\boldsymbol{t}_1, \boldsymbol{t}_2, \ldots, \boldsymbol{t}_m$ for the columns of $\boldsymbol{T}$,

$$\boldsymbol{T}\boldsymbol{u} = u_1\boldsymbol{t}_1 + u_2\boldsymbol{t}_2 + \cdots + u_m\boldsymbol{t}_m,$$

so that the image space $T(U)$ is spanned by the vectors $\boldsymbol{t}_1, \boldsymbol{t}_2, \ldots, \boldsymbol{t}_m$. But the maximum number of linearly independent vectors among the columns $\boldsymbol{t}_1, \boldsymbol{t}_2, \ldots, \boldsymbol{t}_m$ is the column rank of the matrix $\boldsymbol{T}$. It follows that the dimension of the image space $T(U)$ is equal to the column rank of the matrix $\boldsymbol{T}$. Thus the rank of the transformation T is equal to the rank of the matrix $\boldsymbol{T}$.

We can now restate the corollary of Theorem 4.3 as a matrix theorem.

THEOREM 4.6. If the matrix $\boldsymbol{A}$ has rank r, then there exist non-singular matrices $\boldsymbol{H}, \boldsymbol{K}$ such that

$$\boldsymbol{H}^{-1}\boldsymbol{A}\boldsymbol{K} = \begin{bmatrix} \boldsymbol{I}_r & \boldsymbol{0} \\ \boldsymbol{0} & \boldsymbol{0} \end{bmatrix}.$$

THEOREM 4.7. An $n \times n$ matrix $\boldsymbol{A}$ is non-singular if and only if it has rank n.

Proof. (a) Suppose $\boldsymbol{A}$ to be non-singular. Then $\boldsymbol{A}^{-1}$ exists. Let $\boldsymbol{A}$ have rank r. Now $\boldsymbol{A}$ is the matrix of a non-singular linear transformation A on a vector space V of dimension n. The image space $A(V)$ is a subspace of V of dimension r, so that $r \leqslant n$. We now apply the transformation A^{-1} to the space $A(V)$, obtaining the image space $A^{-1}A(V)$. This is a subspace of $A(V)$ of dimension r', say, where $r' \leqslant r$, giving $r' \leqslant r \leqslant n$. But $A^{-1}A(V) = (A^{-1}A)(V) = V$, which has dimension n. Hence $r' = n$, $r = n$.

(b) Suppose A to have rank n. Then the image space has dimension n and so coincides with V, so that every $\boldsymbol{v} \in V$ is the image of some $\boldsymbol{v}_1 \in V$. Suppose that $\boldsymbol{v}$ is also the image of $\boldsymbol{v}_2 \in V$. Then $A(\boldsymbol{v}_1) = \boldsymbol{v} = A(\boldsymbol{v}_2)$, or $A(\boldsymbol{v}_1 - \boldsymbol{v}_2) = \boldsymbol{0}$, so that $\boldsymbol{v}_1 - \boldsymbol{v}_2$ belongs to the kernel of the transformation. However, the dimension of the kernel (or nullity) is given by

$$\begin{aligned} \text{nullity} &= \text{dimension of space} - \text{rank} \\ &= n - n = 0. \end{aligned}$$

The kernel therefore consists of the zero vector alone, and $\boldsymbol{v}_1 - \boldsymbol{v}_2 = \boldsymbol{0}$, $\boldsymbol{v}_1 = \boldsymbol{v}_2$. This shows that every element of V is the image of *exactly one* element of V, and the inverse transformation exists. Hence $\boldsymbol{A}^{-1}$ exists and $\boldsymbol{A}$ is non-singular.

Corollary. The columns (or rows) of any non-singular $n \times n$ matrix form a basis for V.

For, if $\boldsymbol{A}$ is non-singular, its rank is n and its n columns (rows) are therefore linearly independent.

We can use this result to find the coordinates of a given vector in $\boldsymbol{R}_n$ relative to a given basis (see Section 2.8). Let $\{\boldsymbol{a}_1, \boldsymbol{a}_2, \ldots, \boldsymbol{a}_n\}$ be the given basis and let $\boldsymbol{A}$ be the matrix having these vectors as its columns. Then $\boldsymbol{A}$ is non-singular and $\boldsymbol{A}^{-1} = \boldsymbol{B}$ exists and is also non-singular. Hence the rows of $\boldsymbol{B}$, which are denoted by $\{\boldsymbol{b}_1, \boldsymbol{b}_2, \ldots, \boldsymbol{b}_n\}$, are linearly independent and also form a basis of $\boldsymbol{R}_n$. We call it the *reciprocal basis*. Since $\boldsymbol{BA} = \boldsymbol{I}$,

$$\boldsymbol{b}_i\boldsymbol{a}_j = \delta_{ij} \qquad (i = 1, \ldots, n;\ j = 1, \ldots, n),$$

the left-hand side being the inner product of the row vector $\boldsymbol{b}_i$ and the column vector $\boldsymbol{a}_j$. Let $\boldsymbol{u} \in \boldsymbol{R}_n$ be a given vector and let $\boldsymbol{v} = (v_1, v_2, \ldots, v_n)$ be its coordinate vector relative to the basis $\{\boldsymbol{a}_1, \boldsymbol{a}_2, \ldots, \boldsymbol{a}_n\}$. Thus

$$\boldsymbol{u} = v_1\boldsymbol{a}_1 + v_2\boldsymbol{a}_2 + \cdots v_n\boldsymbol{a}_n = \sum_{k=1}^{n} v_k\boldsymbol{a}_k$$

and
$$\boldsymbol{b}_i\boldsymbol{u} = \sum_{k=1}^{n} v_k\boldsymbol{b}_i\boldsymbol{a}_k = v_i \qquad (i = 1, 2, \ldots, n)$$

or
$$\boldsymbol{Bu} = \boldsymbol{v}.$$

Thus the ith coordinate of $\boldsymbol{v}$ relative to the basis consisting of the columns of $\boldsymbol{A}$ is the inner product with $\boldsymbol{v}$ of the ith row of the matrix $\boldsymbol{A}^{-1}$.

In the example of Section 2.8

$$\boldsymbol{A} = \begin{bmatrix} 1 & -1 & 0 \\ 0 & 1 & 2 \\ 2 & 0 & 3 \end{bmatrix},$$

$$\boldsymbol{A}^{-1} = \boldsymbol{B} = \begin{bmatrix} -3 & -3 & 2 \\ -4 & -3 & 2 \\ 2 & 2 & -1 \end{bmatrix},$$

$$\boldsymbol{u} = \{1, -1, 1\}.$$

Hence
$$\boldsymbol{Bu} = \begin{bmatrix} -3 & -3 & 2 \\ -4 & -3 & 2 \\ 2 & 2 & -1 \end{bmatrix} \begin{bmatrix} 1 \\ -1 \\ 1 \end{bmatrix} = \begin{bmatrix} 2 \\ 1 \\ -1 \end{bmatrix}.$$

Similarly $\boldsymbol{Bw} = \{1\ \ 2\ \ 3\}$. Thus the required coordinate vectors are $(2, 1, -1)$ and $(1, 2, 3)$.

THEOREM 4.8. If $\boldsymbol{B}$ is a non-singular square matrix, then $r(\boldsymbol{AB}) = r(\boldsymbol{A})$.

Proof. Let $\boldsymbol{A}$ be of order $n \times m$ and rank r. Then $\boldsymbol{B}$ is of order $m \times m$, and must be of rank m, by Theorem 4.7. $\boldsymbol{B}$ may be regarded as the matrix of a linear mapping B of a vector space V of dimension m. Since B is non-singular, the image space $B(V)$ is of dimension m, and is a subspace of V. Hence $B(V) = V$. $A(V)$ is of dimension r and $A(V) = A\{B(V)\} = (AB)(V)$, so that $(AB)(V)$ is also of dimension r. It follows that $\boldsymbol{AB}$ is of rank r, or $r(\boldsymbol{AB}) = r(\boldsymbol{A})$.

Corollary. If $\boldsymbol{C}$ is a non-singular square matrix, then $r(\boldsymbol{CA}) = r(\boldsymbol{A})$.

For, $r(\boldsymbol{CA}) = r\{(\boldsymbol{CA})^{t}\} = r(\boldsymbol{A}^{t}\boldsymbol{C}^{t}) = r(\boldsymbol{A}^{t})$, since $\boldsymbol{C}^{t}$ is non-singular, and $r(\boldsymbol{A}^{t}) = r(\boldsymbol{A})$.

Example 1. With the notation of Theorem 4.6, find matrices $\boldsymbol{H}$ and $\boldsymbol{K}$ when

$$\boldsymbol{A} = \begin{bmatrix} 1 & 2 & 3 & 4 \\ 0 & 1 & 2 & -1 \\ 2 & 1 & 0 & 11 \end{bmatrix}.$$

We can regard $\boldsymbol{A}$ as the matrix of a linear transformation from $\boldsymbol{R}_4$ into $\boldsymbol{R}_3$, relative to the standard bases in the two spaces. We first find the kernel of the transformation. This is the set of all vectors $\boldsymbol{x} = \{x_1, x_2, x_3, x_4\}$ in $\boldsymbol{R}_4$ such that $\boldsymbol{Ax} = \boldsymbol{0}$. Thus

$$\left.\begin{aligned} x_1 + 2x_2 + 3x_3 + 4x_4 &= 0 \\ x_2 + 2x_3 - x_4 &= 0 \\ 2x_1 + x_2 + 11x_4 &= 0 \end{aligned}\right\}$$

Twice the first equation minus three times the second one gives the third one. Thus there are only two linearly independent equations, e.g., the first two. Taking $x_3 = \lambda$, $x_4 = \mu$ we obtain

$$x_2 = -2\lambda + \mu, \qquad x_1 = \lambda - 6\mu.$$

Hence
$$\boldsymbol{x} = \lambda\{1, -2, 1, 0\} + \mu\{-6, 1, 0, 1\}.$$

Thus the kernel is a two-dimensional space with basis $\{1, -2, 1, 0\}$, $\{-6, 1, 0, 1\}$.

We now extend this to a basis of $\boldsymbol{R}_4$. It is simplest to select two of the standard basis vectors which, together with the two vectors above, form a linearly independent set. In this case we may take $\{1, 0, 0, 0\}$ and $\{0, 1, 0, 0\}$. Now form a matrix $\boldsymbol{K}$ having these four basis vectors of $\boldsymbol{R}_4$ as columns, making the vectors of ker $\boldsymbol{A}$ the *last* two columns. Thus

$$\boldsymbol{K} = \begin{bmatrix} 1 & 0 & 1 & -6 \\ 0 & 1 & -2 & 1 \\ 0 & 0 & 1 & 0 \\ 0 & 0 & 0 & 1 \end{bmatrix}$$

and this is non-singular, since its columns are linearly independent.

$$\boldsymbol{AK} = \begin{bmatrix} 1 & 2 & 3 & 4 \\ 0 & 1 & 2 & -1 \\ 2 & 1 & 0 & 11 \end{bmatrix} \begin{bmatrix} 1 & 0 & 1 & -6 \\ 0 & 1 & -2 & 1 \\ 0 & 0 & 1 & 0 \\ 0 & 0 & 0 & 1 \end{bmatrix} = \begin{bmatrix} 1 & 2 & 0 & 0 \\ 0 & 1 & 0 & 0 \\ 2 & 1 & 0 & 0 \end{bmatrix}.$$

We expect the last two to be zero columns, since the last two columns in $\boldsymbol{K}$ are in ker $\boldsymbol{A}$. The first two columns form a basis for the image space, which is a subspace of $\boldsymbol{R}_3$. Adjoin to these a third column, say $\{0 \quad 0 \quad 1\}$, to form a basis for $\boldsymbol{R}_3$ and let

$$\boldsymbol{H} = \begin{bmatrix} 1 & 2 & 0 \\ 0 & 1 & 0 \\ 2 & 1 & 1 \end{bmatrix}.$$

Then $\boldsymbol{H}$ is non-singular, since its columns are linearly independent, and from $\boldsymbol{H}^{-1}\boldsymbol{H} = \boldsymbol{I}$ it follows that

$$\boldsymbol{H}^{-1}\begin{bmatrix} 1 & 2 \\ 0 & 1 \\ 2 & 1 \end{bmatrix} = \begin{bmatrix} 1 & 0 \\ 0 & 1 \\ 0 & 0 \end{bmatrix}.$$

Thus

$$\boldsymbol{H}^{-1}\boldsymbol{A}\boldsymbol{K} = \begin{bmatrix} \boldsymbol{I}_2 \\ \boldsymbol{0} \end{bmatrix}.$$

4.4 Determinant rank

We can define the rank of a matrix in yet another (equivalent) way, which is sometimes useful in computation.

If $\boldsymbol{A}^{-1}$ exists, then $\det \boldsymbol{A} \neq 0$ (see Section 3.7). Conversely, if $\det \boldsymbol{A} \neq 0$, we can give an explicit formula for $\boldsymbol{A}^{-1}$ as follows. We first calculate the cofactor A_{ij} of each element a_{ij}. This is $(-1)^{i+j}$ multiplied by the determinant obtained from $\det \boldsymbol{A}$ by deleting the ith row and the jth column. We replace each element of $\boldsymbol{A}$ by its cofactor and then transpose the resulting matrix to give the *adjoint* matrix of $\boldsymbol{A}$, denoted by adj $\boldsymbol{A}$. Thus $[\text{adj } \boldsymbol{A}]_{ij} = A_{ji}$.

Using the well-known determinant properties

$$\sum_j a_{ij}A_{ij} = \det \boldsymbol{A} = \sum_i a_{ij}A_{ij},$$

$$\sum_j a_{ij}A_{kj} = 0 \qquad (k \neq i),$$

$$\sum_i a_{ik}A_{ij} = 0 \qquad (k \neq j),$$

we have

$$\boldsymbol{A} \,.\, \text{adj } \boldsymbol{A} = \text{adj } \boldsymbol{A} \,.\, \boldsymbol{A} = \det \boldsymbol{A} \,.\, \boldsymbol{I}.$$

If $\det \boldsymbol{A} \neq 0$ this gives

$$\boldsymbol{A} \,.\, (\det \boldsymbol{A})^{-1} \text{ adj } \boldsymbol{A} = (\det \boldsymbol{A})^{-1} \text{ adj } \boldsymbol{A} \,.\, \boldsymbol{A} = \boldsymbol{I},$$

so that

$$(\det \boldsymbol{A})^{-1} \text{ adj } \boldsymbol{A} = \boldsymbol{A}^{-1}.$$

We have therefore shown that $\boldsymbol{A}^{-1}$ exists if and only if det $\boldsymbol{A} \neq 0$. If det $\boldsymbol{A} \neq 0$, it follows from Theorem 4.7 that $\boldsymbol{A}$ is of rank n and its rows (and also its columns) are linearly independent.

We now define the minors of a matrix $\boldsymbol{A}$, where $\boldsymbol{A}$ is not necessarily square. Obtain a submatrix of $\boldsymbol{A}$ of order $p \times p$ by deleting all the elements of $\boldsymbol{A}$ except those that occur *both* in a specified set of p rows *and* in a specified set of p columns. The determinant of this submatrix is called a *minor of order* p. Let $\boldsymbol{A}$ be a matrix of rank r and order $n \times m$. $\boldsymbol{A}$ has r linearly independent rows and we can therefore delete $n - r$ rows to obtain an $r \times n$ submatrix $\boldsymbol{A}^{(1)}$ whose rows are linearly independent. It then follows that $\boldsymbol{A}^{(1)}$ has r linearly independent columns, and if we delete the remaining $m - r$ columns we obtain a submatrix $\boldsymbol{A}^{(2)}$ of order $r \times r$ whose columns are linearly independent. $\boldsymbol{A}^{(2)}$ then has rank r and is therefore non-singular, so that det $\boldsymbol{A}^{(2)} \neq 0$. We have now shown that, if $\boldsymbol{A}$ is of rank r, then it has a non-zero minor of order r.

Conversely, suppose that $\boldsymbol{A}$ has a non-zero minor of order p. Let $\boldsymbol{B}$ be the corresponding $p \times p$ submatrix of $\boldsymbol{A}$. Then, since det $\boldsymbol{B} \neq 0$, the rows of $\boldsymbol{B}$ are linearly independent and so the same set of p rows of the whole matrix $\boldsymbol{A}$ is certainly linearly independent. The rank of $\boldsymbol{A}$ is therefore not less than p. Combining these results we therefore obtain the following condition.

If the matrix $\boldsymbol{A}$ possesses a non-zero minor of order r, whilst every minor of higher order is zero, then $\boldsymbol{A}$ has rank r.

For matrices of low orders or ones with many zero elements this gives us a quick method of determining rank, but for a matrix of high order it could be very tedious. We therefore develop an alternative method.

4.5 Computation of rank

Consider operations on the rows of matrices which are combinations of three fundamental ones called *elementary row operations*. They are:

(1) the interchange of any two rows;
(2) the multiplication of a row by a non-zero scalar c;
(3) the replacement of the ith row by ith row $+$ jth row.

The result of adding $c \times j$th row to the ith row is a combination of elementary row operations. First, multiply the jth row by c, then add the new jth row to the ith row and finally multiply the new jth row by c^{-1}.

Definition. A matrix $\boldsymbol{B}$ is said to be *row-equivalent* to a matrix $\boldsymbol{A}$ if $\boldsymbol{B}$ can be obtained from $\boldsymbol{A}$ by a succession of elementary row operations.

Each of the elementary operations has an inverse operation which is also an elementary operation or a combination of them. For (1), (2), and (3) the inverses are:

(4) a second interchange of the same two rows;
(5) multiplication of the same row by c^{-1};
(6) a combination of three operations, i.e.,
 (*a*) multiply jth row by -1,
 (*b*) add new jth row to ith row,
 (*c*) multiply jth row by -1.

Hence $\boldsymbol{A}$ can also be obtained from $\boldsymbol{B}$ by a succession of elementary row operations. It follows that, if $\boldsymbol{B}$ is row equivalent to $\boldsymbol{A}$, then $\boldsymbol{A}$ is row-equivalent to $\boldsymbol{B}$, and we write $\boldsymbol{A} \sim \boldsymbol{B}$.

We can now represent each elementary row operation by a matrix. Let $\boldsymbol{A}$ be a matrix of order $n \times m$ and let $\boldsymbol{U}^{(ij)}$ be the matrix obtained from $\boldsymbol{I}_n$ by interchanging its ith and jth rows. Thus

$$u_{ji} = u_{ij} = 1, \qquad u_{jj} = u_{ii} = 0, \qquad u_{hk} = \delta_{hk} \text{ otherwise.}$$

Then

$$[\boldsymbol{U}^{(ij)}\boldsymbol{A}]_{pq} = \sum_{k=1}^{n} u_{pk}a_{kq} = a_{pq} \qquad (p \neq i \text{ or } j),$$

$$[\boldsymbol{U}^{(ij)}\boldsymbol{A}]_{iq} = a_{jq},$$

$$[\boldsymbol{U}^{(ij)}\boldsymbol{A}]_{jq} = a_{iq}.$$

It follows that $\boldsymbol{U}^{(ij)}\boldsymbol{A}$ is the matrix obtained from $\boldsymbol{A}$ by interchanging its ith and jth rows. Moreover,

$$\boldsymbol{U}^{(ij)}\boldsymbol{U}^{(ji)} = \boldsymbol{I}_n,$$

so that $\boldsymbol{U}^{(ij)}$ is non-singular.

Now let $\boldsymbol{C}^{(i)}$ be the $n \times n$ diagonal matrix given by

$$c_{ii} = c \neq 0, \qquad c_{kk} = 1 \quad (k \neq i), \qquad c_{kj} = 0 \quad (k \neq j).$$

Then $\boldsymbol{C}^{(i)}\boldsymbol{A}$ is the matrix obtained from $\boldsymbol{A}$ by multiplying its ith row by c, and $\boldsymbol{C}^{(i)}$ is non-singular, having as its inverse the same matrix in which c is replaced by c^{-1}.

Finally, let $E^{(ij)}$ $(i \neq j)$ be the matrix given by

$$e_{ij} = 1, \qquad e_{hk} = \delta_{hk} \text{ otherwise.}$$

Then $E^{(ij)}A$ is the matrix obtained from A by adding its jth row to its ith row. $E^{(ij)}$ is non-singular, having as its inverse the matrix $F^{(ij)}$ for which

$$f_{ij} = -1, \qquad f_{hk} = \delta_{hk} \text{ otherwise.}$$

Note that $U^{(ij)}$ is the unit matrix with its ith and jth rows interchanged, $C^{(i)}$ is the unit matrix with its ith unit element multiplied by c, and $E^{(ij)}$ is the unit matrix with an additional unit element in the ijth place. These three types $U^{(ij)}$, $C^{(ij)}$, $E^{(ij)}$ are called elementary matrices.

We have shown that each elementary row operation on a matrix A can be effected by premultiplying A by an elementary matrix, and consequently any succession of elementary row operations can be performed by premultiplying A by a product of elementary matrices, all of order $n \times n$, i.e., by a single non-singular $n \times n$ matrix. Thus if $A \sim B$, then there exists a non-singular square matrix P such that $PA = B$. By Theorem 4.7 it follows that row-equivalent matrices have the same rank.

We now construct from a given matrix A a row-equivalent matrix whose rank can be determined by inspection. Let A be of order $n \times m$. If the first column of A contains a non-zero element we can, by interchanging two rows if necessary, replace A by a row-equivalent matrix whose leading element (i.e., the one in the top left-hand corner) is non-zero. By multiplying the first row by a suitable constant we can make this element equal to 1, and then, by subtracting suitable multiples of the first row from the remaining rows in turn, we can make all other elements in the first column equal to zero. In this way we replace A by a row-equivalent matrix of the form

$$\begin{bmatrix} 1 & C_1 \\ 0 & A_1 \end{bmatrix},$$

where A_1 is of order $(n - 1) \times (m - 1)$. On the other hand, if the first column of A consists entirely of zeros, A is of the form

$$\begin{bmatrix} 0 & C_1 \\ 0 & A_1 \end{bmatrix},$$

where A_1 is again of order $(n-1) \times (m-1)$. We now repeat the process with the matrix A_1, obtaining

$$A_1 = \begin{bmatrix} 0 & C_2 \\ \mathbf{0} & A_2 \end{bmatrix} \quad \text{or} \quad A_1 \sim \begin{bmatrix} 1 & C_2 \\ \mathbf{0} & A_2 \end{bmatrix},$$

where A_2 is of order $(n-2) \times (m-2)$.

Continuing in this way there are two possibilities. If $n \geqslant m$, we reach a matrix A_{m-1} of order $(n-m+1) \times 1$. If all elements of A_{m-1} are zeros we have come to the end of the process. If there is at least one non-zero element we can, by elementary row operations, make the leading element 1 and all other elements (if any) zero. On the other hand, if $m > n$, we reach a matrix A_{n-1} of order $1 \times (m-n+1)$, and if the elements of A_{n-1} are not all zero we can make the first non-zero element equal to 1.

We thus arrive at a matrix $\boldsymbol{B}$, row-equivalent to $\boldsymbol{A}$, and having the following structure. The first non-zero element in every row is 1 and the number of zeros preceding this 1 is greater than the corresponding number of zeros in the preceding row. Such a matrix is said to be in *echelon* form. Thus if $\boldsymbol{A}$ is of order 5×8, we may obtain an equivalent matrix $\boldsymbol{B}$ of the form

$$\boldsymbol{B} = \begin{bmatrix} 1 & b_{12} & 0 & b_{14} & b_{15} & b_{16} & b_{17} & b_{18} \\ 0 & 1 & 0 & b_{24} & b_{25} & b_{26} & b_{27} & b_{28} \\ 0 & 0 & 0 & 0 & 1 & b_{36} & b_{37} & b_{38} \\ 0 & 0 & 0 & 0 & 0 & 0 & 1 & b_{48} \\ 0 & 0 & 0 & 0 & 0 & 0 & 0 & 0 \end{bmatrix}.$$

The rank of such a matrix is equal to the number of non-zero rows. For suppose that $\boldsymbol{B}$ has r non-zero rows and form a minor of order r from these r rows by retaining only the r columns containing the unit elements which are the first non-zero elements in their respective rows. In the example above we should form a minor of order 4 by taking the first four rows and columns 1, 2, 5, 7. The leading diagonal elements of such a minor are all equal to 1 and all elements below the diagonal are equal to zero. The minor therefore has the value 1 and is non-zero. On the other hand, every minor of order greater than r has at least one zero row and is therefore equal to zero. The matrix $\boldsymbol{B}$ is consequently of rank r. The matrix of the above example has rank 4.

This method can always be used to determine the rank of a matrix but we can often shorten the process by breaking off at a stage when we can see the determinant rank by inspection. To illustrate both methods, here are two worked examples.

Example 2.

$$\begin{bmatrix} 5 & 9 & 3 \\ -3 & 5 & 6 \\ -1 & -5 & -3 \end{bmatrix} \sim \begin{bmatrix} -1 & -5 & -3 \\ -3 & 5 & 6 \\ 5 & 9 & 3 \end{bmatrix}$$

$$\sim \begin{bmatrix} 1 & 5 & 3 \\ -3 & 5 & 6 \\ 5 & 9 & 3 \end{bmatrix} \sim \begin{bmatrix} 1 & 5 & 3 \\ 0 & 20 & 15 \\ 0 & -16 & -12 \end{bmatrix}$$

$$\sim \begin{bmatrix} 1 & 5 & 3 \\ 0 & 1 & \frac{3}{4} \\ 0 & -16 & -12 \end{bmatrix} \sim \begin{bmatrix} 1 & 5 & 3 \\ 0 & 1 & \frac{3}{4} \\ 0 & 0 & 0 \end{bmatrix}$$

The matrix is of rank 2.

We could, however, have seen the result after the third stage. For the equivalent matrix $\begin{bmatrix} 1 & 5 & 3 \\ 0 & 20 & 15 \\ 0 & -16 & -12 \end{bmatrix}$ has determinant $-20 \times 12 + 15 \times 16 = 0$, but its leading second-order minor $\begin{vmatrix} 1 & 5 \\ 0 & 20 \end{vmatrix} = 20 \neq 0$. The matrix therefore has rank 2.

There is no difficulty in finding short cuts. One can, for example, proceed as follows:

$$\begin{bmatrix} 5 & 9 & 3 \\ -3 & 5 & 6 \\ -1 & -5 & -3 \end{bmatrix} \sim \begin{bmatrix} 1 & 9 & 6 \\ 0 & 32 & 24 \\ 0 & 4 & 3 \end{bmatrix},$$

where the new first row is obtained by adding the second and third rows to the old one and the zeros in the first column are obtained as before. The second matrix clearly has determinant rank 2.

Example 3.

$$\begin{bmatrix} 1 & 2 & 4 & -1 & 5 \\ 1 & 2 & 3 & -1 & 3 \\ -1 & -2 & 0 & 4 & 3 \end{bmatrix} \sim \begin{bmatrix} 1 & 2 & 4 & -1 & 5 \\ 0 & 0 & -1 & 0 & -2 \\ 0 & 0 & 4 & 3 & 8 \end{bmatrix}$$

$$\sim \begin{bmatrix} 1 & 2 & 4 & -1 & 5 \\ 0 & 0 & 1 & 0 & 2 \\ 0 & 0 & 4 & 3 & 8 \end{bmatrix} \sim \begin{bmatrix} 1 & 2 & 4 & -1 & 5 \\ 0 & 0 & 1 & 0 & 2 \\ 0 & 0 & 0 & 3 & 0 \end{bmatrix}$$

$$= \begin{bmatrix} 1 & 2 & 4 & -1 & 5 \\ 0 & 0 & 1 & 0 & 2 \\ 0 & 0 & 0 & 1 & 0 \end{bmatrix}.$$

The matrix is of rank 3.

We could, however, have seen this at once after stage one, for the minor of order 3 formed by the first, third, and fourth columns is non-zero.

4.6 Non-singular matrices

Let $\boldsymbol{A}$ be a non-singular $n \times n$ matrix. Then $\boldsymbol{A}$ has rank n, and by the process described in Section 4.5 we can construct a matrix $\boldsymbol{B}$ of echelon form, row-equivalent to $\boldsymbol{A}$ and thus having n non-zero rows. It follows that $\boldsymbol{B}$ is a triangular matrix, having all its leading diagonal elements equal to 1 and all elements below the diagonal equal to zero. Thus

$$\boldsymbol{B} = \begin{bmatrix} 1 & b_{12} & b_{13} \cdot \cdot \cdot b_{1n} \\ 0 & 1 & b_{23} \cdot \cdot \cdot b_{2n} \\ 0 & 0 & 1 \quad \cdot \cdot \cdot b_{3n} \\ \cdot & \cdot & \cdot \quad \cdot \cdot \cdot \cdot \\ 0 & 0 & 0 \quad \cdot \cdot \cdot 1 \end{bmatrix}.$$

Now, by further row operations, all the elements above the leading diagonal can be reduced to zero. For example, if we subtract $b_{12} \times$ second row from first row, b_{12} is replaced by zero. Now subtract $b_{13} \times$ third row from first row and $b_{23} \times$ third row from second row and b_{13}, b_{23} are both replaced by 0. Hence if $\boldsymbol{A}$ is non-singular, there exists a non-singular matrix $\boldsymbol{P}$ such that $\boldsymbol{PA} = \boldsymbol{I}$. Moreover, $\boldsymbol{P}$ is a product of elementary matrices, so that $\boldsymbol{P}^{-1}$ is also a product of elementary matrices, and $\boldsymbol{A} = \boldsymbol{P}^{-1}$. We have therefore proved the following theorem.

THEOREM 4.9. A square matrix is non-singular if and only if it is row-equivalent to the identity matrix. Moreover, any non-singular square matrix can be expressed as a product of elementary matrices.

Finally we discuss the geometrical interpretation of the last result when $\boldsymbol{A}$ is a non-singular 2×2 matrix. The elementary 2×2 matrices are

$$\boldsymbol{U}^{(12)} = \begin{bmatrix} 0 & 1 \\ 1 & 0 \end{bmatrix}, \quad \boldsymbol{C}^{(1)} = \begin{bmatrix} c & 0 \\ 0 & 1 \end{bmatrix}, \quad \boldsymbol{C}^{(2)} = \begin{bmatrix} 1 & 0 \\ 0 & c \end{bmatrix},$$

$$\boldsymbol{E}^{(12)} = \begin{bmatrix} 1 & 1 \\ 0 & 1 \end{bmatrix}, \quad \boldsymbol{E}^{(21)} = \begin{bmatrix} 1 & 0 \\ 1 & 1 \end{bmatrix}.$$

We regard $\boldsymbol{A}$ as the matrix of a non-singular transformation in the plane, relative to perpendicular unit vectors $\boldsymbol{i}, \boldsymbol{j}$ in the directions of the axes of x, y.

$$\boldsymbol{U}^{(12)}\begin{bmatrix} x \\ y \end{bmatrix} = \begin{bmatrix} 0 & 1 \\ 1 & 0 \end{bmatrix}\begin{bmatrix} x \\ y \end{bmatrix} = \begin{bmatrix} y \\ x \end{bmatrix};$$

$\boldsymbol{U}^{(12)}$ corresponds to a reflection in the line $y = x$.

$$\boldsymbol{C}^{(1)}\begin{bmatrix} x \\ y \end{bmatrix} = \begin{bmatrix} c & 0 \\ 0 & 1 \end{bmatrix}\begin{bmatrix} x \\ y \end{bmatrix} = \begin{bmatrix} cx \\ y \end{bmatrix};$$

$\boldsymbol{C}^{(1)}$ corresponds to a contraction or elongation parallel to the x-axis. Similarly, $\boldsymbol{C}^{(2)}$ corresponds to a contraction or elongation parallel to the y-axis.

$$\boldsymbol{E}^{(12)}\begin{bmatrix} x \\ y \end{bmatrix} = \begin{bmatrix} 1 & 1 \\ 0 & 1 \end{bmatrix}\begin{bmatrix} x \\ y \end{bmatrix} = \begin{bmatrix} x + y \\ y \end{bmatrix};$$

$\boldsymbol{E}^{(12)}$ corresponds to a translation (or shear) parallel to the x-axis. Similarly, $\boldsymbol{E}^{(21)}$ corresponds to a shear parallel to the y-axis.

Hence, by Theorem 4.9, any linear transformation of the plane can be represented as a product of shears, reflections, and contractions (or elongations).

We can obtain this representation as follows. Let $\boldsymbol{A} = \begin{bmatrix} 2 & 6 \\ 3 & -1 \end{bmatrix}$.

$$\begin{bmatrix} \frac{1}{2} & 0 \\ 0 & 1 \end{bmatrix}\boldsymbol{A} = \begin{bmatrix} 1 & 3 \\ 3 & -1 \end{bmatrix}$$

$$\begin{bmatrix} 1 & 0 \\ -3 & 1 \end{bmatrix}\begin{bmatrix} 1 & 3 \\ 3 & -1 \end{bmatrix} = \begin{bmatrix} 1 & 3 \\ 0 & -10 \end{bmatrix}$$

$$\begin{bmatrix} 1 & 0 \\ 0 & -\frac{1}{10} \end{bmatrix}\begin{bmatrix} 1 & 3 \\ 0 & -10 \end{bmatrix} = \begin{bmatrix} 1 & 3 \\ 0 & 1 \end{bmatrix}$$

$$\begin{bmatrix} 1 & -3 \\ 0 & 1 \end{bmatrix}\begin{bmatrix} 1 & 3 \\ 0 & 1 \end{bmatrix} = \begin{bmatrix} 1 & 0 \\ 0 & 1 \end{bmatrix}.$$

Hence $$\begin{bmatrix} 1 & -3 \\ 0 & 1 \end{bmatrix}\begin{bmatrix} 1 & 0 \\ 0 & -\frac{1}{10} \end{bmatrix}\begin{bmatrix} 1 & 0 \\ -3 & 1 \end{bmatrix}\begin{bmatrix} \frac{1}{2} & 0 \\ 0 & 1 \end{bmatrix}\boldsymbol{A} = \begin{bmatrix} 1 & 0 \\ 0 & 1 \end{bmatrix}$$

$$\begin{aligned} \boldsymbol{A} &= \begin{bmatrix} \frac{1}{2} & 0 \\ 0 & 1 \end{bmatrix}^{-1}\begin{bmatrix} 1 & 0 \\ -3 & 1 \end{bmatrix}^{-1}\begin{bmatrix} 1 & 0 \\ 0 & -\frac{1}{10} \end{bmatrix}^{-1}\begin{bmatrix} 1 & -3 \\ 0 & 1 \end{bmatrix}^{-1} \\ &= \begin{bmatrix} 2 & 0 \\ 0 & 1 \end{bmatrix}\begin{bmatrix} 1 & 0 \\ 3 & 1 \end{bmatrix}\begin{bmatrix} 1 & 0 \\ 0 & -10 \end{bmatrix}\begin{bmatrix} 1 & 3 \\ 0 & 1 \end{bmatrix} \\ &= \begin{bmatrix} 2 & 0 \\ 0 & 1 \end{bmatrix}\begin{bmatrix} 1 & 0 \\ 1 & 1 \end{bmatrix}^3\begin{bmatrix} 1 & 0 \\ 0 & -10 \end{bmatrix}\begin{bmatrix} 1 & 1 \\ 0 & 1 \end{bmatrix}^3. \end{aligned}$$

$\boldsymbol{A}$ is therefore equivalent to a succession of elementary operations.

Problems

4.1 Find the rank of the matrix

$$\begin{bmatrix} 1 & 1 & 1 & 1 \\ 1 & x-2 & 3 & -2 \\ 2 & 3x-7 & 2x-6 & -x \\ 3 & 2x-1 & x & -3 \end{bmatrix}$$

for all values of the real number x.

4.2 V is a finite-dimensional vector space over the field F and $T \in L(V, F)$. Prove that the rank of T is either 0 or 1.

4.3 T is a linear mapping on the finite-dimensional vector space V. Prove that

$$\text{ket } T \subset \text{ket } T^2 \qquad \text{and} \qquad T(U) \supset T^2(U).$$

4.4 Find non-singular matrices $\boldsymbol{H}$ and $\boldsymbol{K}$ such that $\boldsymbol{H}^{-1}\boldsymbol{A}\boldsymbol{K}$ is in the normal form $\begin{bmatrix} \boldsymbol{I}_r & \boldsymbol{0} \\ \boldsymbol{0} & \boldsymbol{0} \end{bmatrix}$ when

$$(a)\quad \boldsymbol{A} = \begin{bmatrix} 1 & 1 & 2 \\ -1 & -3 & 8 \\ 4 & -3 & -7 \\ 1 & 12 & -3 \end{bmatrix}, \qquad (b)\quad \boldsymbol{A} = \begin{bmatrix} 1 & 2 & 3 & 4 \\ 0 & 1 & -1 & 5 \\ 3 & 4 & 11 & 2 \end{bmatrix}.$$

4.5 Reduce to echelon form and determine the rank of the following matrices;

$$(a)\quad \begin{bmatrix} 2 & -3 & -1 & 1 \\ 3 & 4 & -4 & -3 \\ 0 & 17 & -5 & -9 \end{bmatrix}, \text{ with elements in } \boldsymbol{R};$$

$$(b)\quad \begin{bmatrix} 1 & i & 1+i & 2-i \\ i & 1 & 3 & 2+i \\ 0 & 1-i & 5 & 1 \\ 2i & -1+i & -3+i & 2+3i \end{bmatrix}, \text{ with elements in } \boldsymbol{C}.$$

4.6 Determine the rank of the matrix $\begin{bmatrix} 4 & 3 & 1 \\ 0 & 1 & 0 \\ 2 & 0 & 3 \end{bmatrix}$

(*a*) when the coefficients are in $\boldsymbol{R}$,
(*b*) when the coefficients are in $\boldsymbol{Z}_5$.

4.7 The linear transformation $T \in L(\boldsymbol{R}_3, \boldsymbol{R}_2)$ is given by $T(x_1, x_2, x_3) = (x_1, x_2)$. Find the kernel and image spaces of T, and verify the dimension formula of Theorem 4.2.

4.8 T is a linear transformation in $L(U, V)$ and $\boldsymbol{u}_1, \boldsymbol{u}_2, \ldots, \boldsymbol{u}_n \in U$. Given that $T\boldsymbol{u}_1, T\boldsymbol{u}_2, \ldots, T\boldsymbol{u}_n$ are linearly independent elements of

V, prove that $\boldsymbol{u}_1, \boldsymbol{u}_2, \ldots, \boldsymbol{u}_n$ are linearly independent elements of U. Give an example to show that the converse is false.

4.9 P is the vector space formed by all polynomials p in x, with real coefficients, of degree $\leqslant n$. S is the linear mapping defined by $S(p) = p'$, the derivative of p with respect to x. Find the kernel and image spaces of S and verify the dimension formula of Theorem 4.2.

4.10 V_1, V_2, V_3 are vector spaces over the field F. $T_1 \in L(V_1, V_2)$ and $T_2 \in L(V_2, V_3)$, where $T_2(V_2) = V_3$ and $\ker T_2 = T_1(V_1)$. Prove that

$$\dim V_1 - \dim V_2 + \dim V_3 = 0.$$

5 · *Linear Equations*

We now turn to the solution of linear equations. Using the concept of rank and nullity we are at once able to determine whether a system of equations has a solution and, if it has, precisely how many solutions it possesses.

5.1 Homogeneous equations

Consider the set of m homogeneous linear equations in n unknowns $x_1, x_2, \ldots, x_n$

$$\begin{aligned}
a_{11}x_1 + a_{12}x_2 + \cdots + a_{1n}x_n &= 0, \\
a_{21}x_1 + a_{22}x_2 + \cdots + a_{2n}x_n &= 0, \\
\cdots\cdots\cdots\cdots\cdots\cdots\cdots & \\
a_{m1}x_1 + a_{m2}x_2 + \cdots + a_{mn}x_n &= 0.
\end{aligned}$$

The equations are called *homogeneous* because the constant term in each one is zero, so that every term is of the first degree in the x's. These equations can be written in matrix form

$$\boldsymbol{Ax} = \boldsymbol{0}$$

where $\boldsymbol{A}$ is the $m \times n$ matrix formed by the coefficients, $\boldsymbol{x}$ is the column vector $\{x_1, x_2, \ldots, x_n\}$, and the zero on the right is a zero column vector of order m. The equations always have the solution $\boldsymbol{x} = \boldsymbol{0}$, i.e., $x_1 = x_2 = \cdots = x_n = 0$, but this is the *trivial* solution. We are interested in *non-trivial* (i.e., non-zero) solutions, when they exist. If $\boldsymbol{x}$ is such a solution then, since $\boldsymbol{A}(\lambda\boldsymbol{x}) = \lambda\boldsymbol{Ax} = \boldsymbol{0}$ whenever $\boldsymbol{Ax} = \boldsymbol{0}$, it follows that $\lambda\boldsymbol{x}$ is also a solution for every scalar λ, so that a non-trivial solution is never unique.

Now $\boldsymbol{Ax} = \boldsymbol{0}$ if and only if $\boldsymbol{x}$ belongs to the kernel of the linear transformation in $L(V_n, V_m)$ having matrix $\boldsymbol{A}$ relative to the standard bases. Thus the solutions of the equation form a vector space of dimension equal to the dimension of the kernel of $\boldsymbol{A}$, i.e., equal to the nullity of $\boldsymbol{A}$. But, by Theorem 4.2,

$$\text{nullity of } \boldsymbol{A} = n - r,$$

where r is the rank of $\boldsymbol{A}$. We therefore have the following theorem.

THEOREM 5.1. The set of m homogeneous equations in n unknowns $\boldsymbol{Ax} = \boldsymbol{0}$, where $\boldsymbol{A}$ is an $m \times n$ matrix of rank r, has $n - r$ linearly independent solutions.

If $r = n$, $n - r = 0$. The solution space then has dimension zero and consists of the zero vector alone, so that the equations have only the trivial solution.

Corollary 1. If $m < n$, the equations have a non-trivial solution.

For, $r \leqslant m$ so that $r < n$ in this case and $n - r$ is positive. Hence, if there are fewer equations than unknowns, there is always a non-trivial solution.

Corollary 2. If $m = n$ the equations have non-trivial solutions if and only if $\boldsymbol{A}$ is singular.

For, $\boldsymbol{A}$ is of order $n \times n$ and non-trivial solutions exist if and only if $n - r > 0$, i.e., $r < n$. By Theorem 4.7, $r < n$ if and only if $\boldsymbol{A}$ is singular.

We can also see the necessity of the condition in another way. If $\boldsymbol{A}$ is square and non-singular, then $\boldsymbol{A}^{-1}$ exists, so that if

$$\boldsymbol{Ax} = \boldsymbol{0},$$

then

$$(\boldsymbol{A}^{-1})\boldsymbol{Ax} = \boldsymbol{0},$$

$$(\boldsymbol{A}^{-1}\boldsymbol{A})\boldsymbol{x} = \boldsymbol{0},$$

$$\boldsymbol{x} = \boldsymbol{0}.$$

The equations then have only the trivial solution.

Returning to the general case, if $\boldsymbol{A}$ is of order $m \times n$ and rank r, then $\boldsymbol{A}$ has r linearly independent columns. Let $\boldsymbol{a}_1, \boldsymbol{a}_2, \ldots, \boldsymbol{a}_n$ be the columns of $\boldsymbol{A}$ and let $\boldsymbol{a}_{k_1}, \boldsymbol{a}_{k_2}, \ldots, \boldsymbol{a}_{k_r}$ be r linearly independent columns. Let $\boldsymbol{a}_{h_1}, \boldsymbol{a}_{h_2}, \ldots, \boldsymbol{a}_{h_s}$ be the remaining $n - r$ columns, so that $r + s = n$ and $(k_1, \ldots, k_r, h_1, \ldots, h_s)$ is simply a permutation of the first n positive integers. The equation $\boldsymbol{Ax} = \boldsymbol{0}$ can be written

$$x_1\boldsymbol{a}_1 + x_2\boldsymbol{a}_2 + \cdots + x_n\boldsymbol{a}_n = \boldsymbol{0}. \tag{1}$$

We can choose the values of the $n - r$ unknowns $x_{h_1}, x_{h_2}, \ldots, x_{h_s}$ arbitrarily. But $\boldsymbol{a}_{h_1}, \boldsymbol{a}_{h_2}, \ldots, \boldsymbol{a}_{h_s}$ can be expressed uniquely as linear combinations of $\boldsymbol{a}_{k_1}, \boldsymbol{a}_{k_2}, \ldots, \boldsymbol{a}_{k_r}$ so that

$$x_{h_1}\boldsymbol{a}_{h_1} + x_{h_2}\boldsymbol{a}_{h_2} + \cdots + x_{h_s}\boldsymbol{a}_{h_s} = y_1\boldsymbol{a}_{k_1} + y_2\boldsymbol{a}_{k_2} + \cdots + y_r\boldsymbol{a}_{k_r},$$

where $y_1, y_2, \ldots, y_r$ are expressed in terms of $x_{h_1}, x_{h_2}, \ldots, x_{h_s}$. Equation (1) then becomes

$$(x_{k_1} + y_1)\boldsymbol{a}_{k_1} + (x_{k_2} + y_2)\boldsymbol{a}_{k_2} + \cdots + (x_{k_r} + y_r)\boldsymbol{a}_{k_r} = \boldsymbol{0}.$$

Since $\boldsymbol{a}_{k_1}, \boldsymbol{a}_{k_2}, \ldots, \boldsymbol{a}_{k_r}$ are linearly independent this gives

$$x_{k_1} = -y_1, \qquad x_{k_2} = -y_2, \qquad \ldots, \qquad x_{k_r} = -y_r.$$

Thus the values of $n - r$ of the unknowns (suitably chosen) may be assigned arbitrarily and the values of the remaining r unknowns are then uniquely determined by them.

Example 1. Discuss the solution of the equations

$$\begin{aligned} x + y + z &= 0, \\ x + y - kz &= 0, \\ kx - y - z &= 0, \\ x - ky + z &= 0, \end{aligned}$$

the field of scalars being the real numbers $\boldsymbol{R}$.

The matrix of the equations is

$$\boldsymbol{A} = \begin{bmatrix} 1 & 1 & 1 \\ 1 & 1 & -k \\ k & -1 & -1 \\ 1 & -k & 1 \end{bmatrix}.$$

If $k = -1$, rows 1, 2, and 4 become identical and equal to minus the third row. In this case $\boldsymbol{A}$ has only *one* linearly independent row, the rank of $\boldsymbol{A}$ is 1, and $n - r = 3 - 1 = 2$. The equations therefore have two linearly independent solutions. Since the columns of A are identical in this case, any two of the unknowns may be chosen arbitrarily, say x and y. z is then determined by the equation $x + y + z = 0$. Thus all solutions of the equation have the form

$$x = h, \qquad y = k, \qquad z = -(h + k).$$

Taking $h = 0$, $k = 1$ and $h = 1$, $k = 0$ in turn, we see that $\{0, 1, -1\}$ and $\{1, 0, -1\}$ are both solutions. They are clearly linearly independent and every other solution is of the form $k\{0, 1, -1\} + h\{1, 0, -1\}$. If $k \neq -1$, the minor formed by the first three rows of A is

$$\begin{vmatrix} 1 & 1 & 1 \\ 1 & 1 & -k \\ k & -1 & -1 \end{vmatrix} = \begin{vmatrix} 0 & 0 & 1+k \\ 1 & 1 & -k \\ k & -1 & -1 \end{vmatrix} = -(1 + k)^2 \neq 0.$$

A is therefore of rank 3 and the equations have no non-trivial solution.

The next example illustrates the dependence of the solutions upon the field of scalars.

Example 2. A language school offers three types of course, A, B, and C. Each course is to consist of a certain number of 1-hour lectures per week and fees for each language are charged for multiples of 5 hours so that, for example, 13 hours is charged as 15 hours. In any one language a student has a choice of three combinations of courses: $1A$, $1C$, and $2B$; $2A$ and $1C$; $2A$ and $1B$. Each student must study either two or three languages and no student may attend lectures for more than 15 hours per week. How many hours per week should be devoted to each A, B, C course respectively?

Let x, y, z hours per week be spent on each A, B, C course respectively. It would be most economical for the students if their total number of lecture hours could always be a multiple of 5. We should like, therefore, to determine x, y, and z so that $x + 2y + z$, $2x + z$, $2x + y$ are all multiples of 5, i.e., are all equal to zero (mod 5). We choose the field $\boldsymbol{Z}_5$ of integers modulo 5 as our field of scalars and the equations then have the form

$$\begin{aligned} x + 2y + z &= 0, \\ 2x \qquad + z &= 0, \\ 2x + y \qquad &= 0. \end{aligned}$$

The matrix, A, of these equations is $\begin{bmatrix} 1 & 2 & 1 \\ 2 & 0 & 1 \\ 2 & 1 & 0 \end{bmatrix}$. Now

$$\det A = 2\begin{vmatrix} 2 & 1 \\ 0 & 1 \end{vmatrix} - \begin{vmatrix} 1 & 1 \\ 2 & 1 \end{vmatrix} = 5 = 0 \pmod 5.$$

But

$$\begin{vmatrix} 1 & 2 \\ 2 & 0 \end{vmatrix} = -4 \not\equiv 0 \pmod 5.$$

Hence A is of rank 2 and the equations have $3 - 2 = 1$ linearly independent solution.

The first two columns of A are linearly independent, so that we can choose z arbitrarily and x, y are then determined in terms of z. Put $z = k$ ($k = 0, 1, 2, 3, 4$). Then, since $2x + z = 0$,

$$2x = -k = 4k \pmod 5,$$

$$x = 2k \pmod 5,$$

$$y = -2x = k \pmod 5.$$

Hence $x = 2k$, $y = k$, $z = k$ is a solution and all solutions are of this form.

$k = 1$ gives $x = 2$, $y = 1$, $z = 1$. If A, B, C courses require 2, 1, 1 hours per week respectively, a student taking three languages has 15 hours per week, which is the maximum permitted.

$k = 2$, giving $x = 4$, $y = 2$, $z = 2$ makes it impossible for any student to take three languages.

$k = 3$, giving $x = 6 = 1 \pmod 5$, $y = 3$, $z = 3$ is a possibility, provided that a student taking three languages chooses only two types of course in each language.

$k = 4$, giving $x = 8 = 3 \pmod 5$, $y = 4$, $z = 4$ makes it impossible for any student to take more than one language.

We thus have two possibilities:

$$x = 2, \qquad y = 1, \qquad z = 1;$$

$$x = 1, \qquad y = 3, \qquad z = 3.$$

Note that these solutions are *not* linearly independent, the second being three times the first.

We could easily have obtained these solutions by trial and error, so the above procedure may seem unnecessary. However, suppose that, instead of the given conditions, we had been given

$$\begin{aligned} x + 2y + z &= 0, \\ x \phantom{{}+2y} + 2z &= 0, \\ 2x + y \phantom{{}+2z} &= 0. \end{aligned}$$

The matrix of these equations is

$$A = \begin{bmatrix} 1 & 2 & 1 \\ 1 & 0 & 2 \\ 2 & 1 & 0 \end{bmatrix}$$

and $$\det A = 2\begin{vmatrix} 2 & 1 \\ 0 & 2 \end{vmatrix} - \begin{vmatrix} 1 & 1 \\ 1 & 2 \end{vmatrix} = 7 = 2 \pmod 5 \neq 0.$$

In this case the equations have *no* non-trivial solution. It might have taken a long time to reach this conclusion by trial and error. Again, if the field had been $\boldsymbol{Z}_{37}$ and the number of unknowns greater, trial and error could be a very lengthy procedure. This simple example illustrates the type of problem for which fields other than $\boldsymbol{R}$ and $\boldsymbol{C}$ are required.

Finally, consider the same set of equations

$$\begin{aligned} x + 2y + z &= 0, \\ 2x \phantom{{}+2y} + z &= 0, \\ 2x + y \phantom{{}+z} &= 0, \end{aligned}$$

but with the real numbers $\boldsymbol{R}$ as the field of scalars. In this case $\det \boldsymbol{A} = 5 \neq 0$, the matrix $\boldsymbol{A}$ has rank 3, and the equations have *no* non-trivial solution. Thus the *same* set of equations may have solutions in one field but not in another.

5.2 Non-homogeneous equations

Now consider the set of m non-homogeneous linear equations in n unknowns $x_1, x_2, \ldots, x_n$

$$\begin{aligned} a_{11}x_1 + a_{12}x_2 + \cdots + a_{1n}x_n &= b_1, \\ a_{21}x_1 + a_{22}x_2 + \cdots + a_{2n}x_n &= b_2, \\ \cdots\cdots\cdots\cdots\cdots\cdots\cdots\cdots \\ a_{m1}x_1 + a_{m2}x_2 + \cdots + a_{mn}x_n &= b_m, \end{aligned}$$

where the b's are not all zero. In matrix form the equations are

$$\boldsymbol{A}\boldsymbol{x} = \boldsymbol{b},$$

where $\boldsymbol{A}$ is the $m \times n$ matrix formed by the coefficients, $\boldsymbol{x} = \{x_1\, x_2 \ldots x_n\}$, $\boldsymbol{b} = \{b_1\, b_2 \ldots b_m\}$. As before, $\boldsymbol{A}$ is the matrix of a linear transformation in $L(V_n, V_m)$.

THEOREM 5.2. Let the $m \times n$ matrix $\boldsymbol{A}$ be of rank r. Then the equations $\boldsymbol{A}\boldsymbol{x} = \boldsymbol{b}$ have a solution $\boldsymbol{x}$ if and only if $\boldsymbol{b}$ belongs to the r-dimensional subspace of V spanned by the columns of $\boldsymbol{A}$. All solutions of the equation can be obtained by adding to any one solution an arbitrary solution of the homogeneous equation $\boldsymbol{A}\boldsymbol{x} = \boldsymbol{0}$.

Proof. The set of all vectors $\boldsymbol{A}\boldsymbol{x}$, where $\boldsymbol{x} \in V_n$, is the image space of $\boldsymbol{A}$ and has dimension r. But, with the notation used in Theorem 5.1,

$$\boldsymbol{A}\boldsymbol{x} = x_1\boldsymbol{a}_1 + x_2\boldsymbol{a}_2 + \cdots + x_n\boldsymbol{a}_n,$$

so that $\{\boldsymbol{a}_1, \boldsymbol{a}_2, \ldots, \boldsymbol{a}_n\}$ is a spanning set for the image space. If $\boldsymbol{A}\boldsymbol{x} = \boldsymbol{b}$ has a solution $\boldsymbol{x}$, it follows that $\boldsymbol{b}$ must belong to the image space of $\boldsymbol{A}$, i.e., to the subspace of V_m spanned by the columns of $\boldsymbol{A}$.

Conversely, suppose that $\boldsymbol{b}$ does belong to the subspace spanned by $\{\boldsymbol{a}_1, \boldsymbol{a}_2, \ldots, \boldsymbol{a}_n\}$. Then $\boldsymbol{b}$ can be expressed as a linear combination of $\{\boldsymbol{a}_1, \boldsymbol{a}_2, \ldots, \boldsymbol{a}_n\}$ so that

$$\boldsymbol{b} = x_1\boldsymbol{a}_1 + x_2\boldsymbol{a}_2 + \cdots + x_n\boldsymbol{a}_n = \boldsymbol{A}\boldsymbol{x}$$

and the equation $\boldsymbol{A}\boldsymbol{x} = \boldsymbol{b}$ has a solution.

Let $\boldsymbol{z}$ be a given solution and $\boldsymbol{y}$ another solution of the equation, so that $\boldsymbol{A}\boldsymbol{y} = \boldsymbol{b}$, $\boldsymbol{A}\boldsymbol{z} = \boldsymbol{b}$. Then $\boldsymbol{A}(\boldsymbol{y} - \boldsymbol{z}) = \boldsymbol{0}$, and $\boldsymbol{y} - \boldsymbol{z}$ is a solution of the homogeneous equation $\boldsymbol{A}\boldsymbol{x} = \boldsymbol{0}$. Thus $\boldsymbol{y} = \boldsymbol{z} + \xi$, where $\boldsymbol{A}\xi = \boldsymbol{0}$. Conversely, if $\boldsymbol{A}\boldsymbol{z} = \boldsymbol{b}$ and $\boldsymbol{A}\xi = \boldsymbol{0}$, then $\boldsymbol{A}(\boldsymbol{z} + \xi) = \boldsymbol{b} + \boldsymbol{0} = \boldsymbol{b}$, so that $\boldsymbol{z} + \xi$ is a solution of the given equation.

$\boldsymbol{A}$ has r linearly independent columns. As in Theorem 5.1 we can choose arbitrarily the values of the variables x corresponding to the remaining $n - r$ columns of $\boldsymbol{A}$. The x's corresponding to the r linearly independent columns are then determined uniquely by the equations.

Definition. The matrix $[\boldsymbol{A}\ \boldsymbol{b}]$ of order $m \times (n + 1)$ obtained by adjoining to $\boldsymbol{A}$ the additional column vector $\boldsymbol{b}$ is called the *augmented matrix* of the equations.

We can now restate Theorem 5.2 in matrix form as follows.

The equations $\boldsymbol{A}\boldsymbol{x} = \boldsymbol{b}$ have a solution if and only if the matrix $\boldsymbol{A}$ and the augmented matrix $[\boldsymbol{A}\ \boldsymbol{b}]$ have the same rank.

In the particular case when the number of equations is equal to the number of unknowns, $\boldsymbol{A}$ is square of order $n \times n$. If $\boldsymbol{A}$ is non-singular, $\boldsymbol{A}^{-1}$ exists, and $\boldsymbol{Ax} = \boldsymbol{b}$ implies

$$\boldsymbol{A}^{-1}(\boldsymbol{Ax}) = \boldsymbol{A}^{-1}\boldsymbol{b},$$

i.e.,

$$(\boldsymbol{A}^{-1}\boldsymbol{A})\boldsymbol{x} = \boldsymbol{A}^{-1}\boldsymbol{b},$$

or

$$\boldsymbol{x} = \boldsymbol{A}^{-1}\boldsymbol{b}.$$

Here the equations have the unique solution $\boldsymbol{A}^{-1}\boldsymbol{b}$. This agrees with the above result, for $\boldsymbol{A}$ has rank n and $[\boldsymbol{A}\ \boldsymbol{b}]$, whose rank cannot exceed n, must also be of rank n.

We can often simplify the solution of a set of equations by reducing the matrix $\boldsymbol{A}$ to triangular or echelon form. We can find a non-singular matrix $\boldsymbol{P}$ (equivalent to a succession of elementary row operations) such that $\boldsymbol{PA} = \boldsymbol{B}$, where $\boldsymbol{B}$ is in echelon form. The equation $\boldsymbol{Ax} = \boldsymbol{b}$ is then equivalent to $\boldsymbol{PAx} = \boldsymbol{Pb}$, where $\boldsymbol{Pb}$ is obtained from $\boldsymbol{b}$ by performing on it the same succession of row operations. The new equations can now be solved in turn, beginning with the last one and working backwards.

Example 3. Find the condition which must be satisfied by y_1, y_2, y_3, y_4 in order that the equations

$$\begin{aligned} x_1 - x_3 + 3x_4 + x_5 &= y_1, \\ 2x_1 + x_2 - 2x_4 - x_5 &= y_2, \\ x_1 + 2x_2 + 2x_3 + 4x_5 &= y_3, \\ x_2 + x_3 + 5x_4 + 6x_5 &= y_4 \end{aligned}$$

shall have a solution x_1, x_2, x_3, x_4, x_5. Find all the solutions for $y_1 = -3$, $y_2 = 5$, $y_3 = 6$, $y_4 = -2$.

The equations are

$$\begin{bmatrix} 1 & 0 & -1 & 3 & 1 \\ 2 & 1 & 0 & -2 & -1 \\ 1 & 2 & 2 & 0 & 4 \\ 0 & 1 & 1 & 5 & 6 \end{bmatrix} \begin{bmatrix} x_1 \\ x_2 \\ x_3 \\ x_4 \\ x_5 \end{bmatrix} = \begin{bmatrix} y_1 \\ y_2 \\ y_3 \\ y_4 \end{bmatrix}.$$

Now the last row of the 4×5 matrix is equal to the sum of the first and third rows minus the second. Performing the appropriate row operations on both sides of the equation we can reduce the last row of the 4×5 matrix to zero, obtaining the equations

$$\begin{bmatrix} 1 & 0 & -1 & 3 & 1 \\ 2 & 1 & 0 & -2 & -1 \\ 1 & 2 & 2 & 0 & 4 \\ 0 & 0 & 0 & 0 & 0 \end{bmatrix} \begin{bmatrix} x_1 \\ x_2 \\ x_3 \\ x_4 \\ x_5 \end{bmatrix} = \begin{bmatrix} y_1 \\ y_2 \\ y_3 \\ y_4 - y_3 + y_2 - y_1 \end{bmatrix}.$$

The last of these four equations gives

$$y_4 - y_3 + y_2 - y_1 = 0,$$

so that the equations have no solution if this condition is not satisfied. If it is, the first three equations can then be written

$$\begin{bmatrix} 1 & 0 & -1 & 3 & 1 \\ 2 & 1 & 0 & -2 & -1 \\ 1 & 2 & 2 & 0 & 4 \end{bmatrix} \begin{bmatrix} x_1 \\ x_2 \\ x_3 \\ x_4 \\ x_5 \end{bmatrix} = \begin{bmatrix} y_1 \\ y_2 \\ y_3 \end{bmatrix}.$$

Elementary row operations reduce this to

$$\begin{bmatrix} 1 & 0 & -1 & 3 & 1 \\ 0 & 1 & 2 & -8 & -3 \\ 0 & 2 & 3 & -3 & 3 \end{bmatrix} \begin{bmatrix} x_1 \\ x_2 \\ x_3 \\ x_4 \\ x_5 \end{bmatrix} = \begin{bmatrix} y_1 \\ y_2 - 2y_1 \\ y_3 - y_1 \end{bmatrix},$$

from which we obtain

$$\begin{bmatrix} 1 & 0 & -1 & 3 & 1 \\ 0 & 1 & 2 & -8 & -3 \\ 0 & 0 & -1 & 13 & 9 \end{bmatrix} \begin{bmatrix} x_1 \\ x_2 \\ x_3 \\ x_4 \\ x_5 \end{bmatrix} = \begin{bmatrix} y_1 \\ y_2 - 2y_1 \\ (y_3 - y_1) - 2(y_2 - 2y_1) \end{bmatrix}$$

$$= \begin{bmatrix} y_1 \\ y_2 - 2y_1 \\ 3y_1 - 2y_2 + y_3 \end{bmatrix}.$$

The 3×5 matrix has rank 3 and its first three columns are linearly independent so that we may choose x_4 and x_5 arbitrarily. Take $x_4 = h$, $x_5 = k$. The equations in reverse order are then

$$\begin{aligned} -x_3 + 13h + 9k &= 3y_1 - 2y_2 + y_3, \\ x_2 + 2x_3 - 8h - 3k &= y_2 - 2y_1, \\ x_1 - x_3 + 3h + k &= y_1. \end{aligned}$$

These determine x_3, x_2, and x_1 in turn, in terms of h and k. For the given values of y_1, y_2, y_3, y_4

$$y_4 - y_3 + y_2 - y_1 = -2 - 6 + 5 + 3 = 0.$$

Hence the equations are consistent for these values of y and the solution is given by

$$\begin{aligned} x_4 &= h, \qquad x_5 = k, \\ x_3 &= 13h + 9k + 9 + 10 - 6 = 13h + 9k + 13, \\ x_2 &= -26h - 18k - 26 + 8h + 3k + 5 + 6 \\ &= -18h - 15k - 15, \\ x_1 &= 13h + 9k + 13 - 3h - k - 3 \\ &= 10h + 8k + 10. \end{aligned}$$

Thus all the solutions of the equation are given by

$$\boldsymbol{x} = \{10h + 8k + 10, -18h - 15k - 15, 13h + 9k + 13, h, k\},$$

where h and k are arbitrary.

Problems

5.1 Find all the solutions of the equations

$$\begin{aligned} 2x_1 - 3x_2 - x_3 + x_4 &= 0, \\ 3x_1 + 4x_2 - 4x_3 - 3x_4 &= 0, \\ 17x_2 - 5x_3 - 9x_4 &= 0. \end{aligned}$$

Show that there is one solution that also satisfies the equations

$$\begin{aligned} x_1 + x_2 + x_3 + x_4 + 1 &= 0, \\ x_1 - x_2 - x_3 - x_4 - 3 &= 0. \end{aligned}$$

but that there is no solution that is linearly dependent on the pair $\{0, 1, 2, 3\}$, $\{3, 2, 1, 0\}$.

5.2 Solve the equations

$$\begin{aligned} 2x + y + 3z &= 0, \\ 3x - 2y + z &= 0, \\ x - 3y - 2z &= 0. \end{aligned}$$

5.3 Solve the equations

$$\begin{aligned} 2x - y - z &= 0, \\ x + y + 2z &= 0, \\ 7x + y - 3z &= 0, \\ 2y - z &= 0. \end{aligned}$$

5.4 Solve the equations

$$\begin{aligned} x + y + 2z + w &= 5, \\ 3x + 2y - z + 3w &= 6, \\ 4x + 3y + z + 4w &= 11, \\ 2x + y - 3z + 2w &= 1. \end{aligned}$$

5.5 Find all rational solutions of the system of equations

$$\begin{aligned} x + y - z - w &= -1, \\ 3x + 4y - z - 2w &= 3, \\ x + 2y + z &= 5. \end{aligned}$$

5.6 Prove that the equations

$$\begin{aligned} x_1 + 2x_2 + 3x_3 - 3x_4 &= k_1 \\ 2x_1 - 5x_2 - 3x_3 + 12x_4 &= k_2 \\ 7x_1 + x_2 + 8x_3 + 5x_4 &= k_3 \end{aligned}$$

have a solution in $\boldsymbol{R}_4$ if and only if $37k_1 + 13k_2 - 9k_3 = 0$. Find all the solutions when $k_1 = 1$, $k_2 = 2$, $k_3 = 7$.

5.7 Find all the solutions of the following system of linear equations;

$$\begin{aligned} 7x + y - z &= 0, \\ 2x - y + z - 3k &= 0, \\ x + y - z + 2k &= 0. \end{aligned}$$

5.8 Find the value of λ for which the equations

$$\begin{aligned} x + \lambda y - z &= 1 \\ 2x + y + 2z &= 5\lambda + 1 \\ x - y + 3z &= 4\lambda + 2 \\ x - 2\lambda y + 7z &= 10\lambda - 1 \end{aligned}$$

are consistent and find all the solutions in this case.

5.9 Find the dimension of the space of solutions of the following systems of linear equations and in each case find a basis for the space.

$$\begin{aligned} \text{(a)}\quad x + 2y - z &= 0, \\ x - y &= 0, \\ \text{(b)}\quad 3x - 2y + z &= 0. \end{aligned}$$

6 · *Eigenvalues and eigenvectors*

In Chapter 4 we have shown that, given any linear transformation $T \in L(U, V)$, where U and V are distinct spaces, we can always choose bases for U and V relative to which the matrix of T is diagonal. The argument used there does not necessarily hold if $U = V$, i.e., if T is a transformation on the vector space V. We should still like to know whether it is possible to find a basis for V relative to which T has a diagonal matrix. Unfortunately this cannot always be done and we now investigate the problem. It is closely bound up with the fundamental problem of determining whether vectors $\boldsymbol{v} \neq \boldsymbol{0}$ exist such that $T(\boldsymbol{v}) = \lambda \boldsymbol{v}$, λ being a scalar. In three dimensions a non-zero vector $\boldsymbol{v}$ determines a direction (all vectors $c\boldsymbol{v}$ having the same direction) and we are looking for a direction that is unchanged by the transformation. If $T(\boldsymbol{v}) = \lambda \boldsymbol{v}$, then $T(c\boldsymbol{v}) = \lambda c \boldsymbol{v}$ for every scalar c, so that the one-dimensional subspace generated by (i.e., spanned by) $\boldsymbol{v}$ is mapped into itself by T.

Definition. A vector $\boldsymbol{v} \neq \boldsymbol{0}$ such that $T(\boldsymbol{v}) = \lambda \boldsymbol{v}$ for some scalar λ is called an *eigenvector* (characteristic vector) of T. A scalar λ such that $T(\boldsymbol{v}) = \lambda \boldsymbol{v}$ for some non-zero vector $\boldsymbol{v}$ is called an *eigenvalue* (characteristic value) of T. It is important to note that, whilst λ may take the value zero, $\boldsymbol{v}$ is non-zero by definition. Similarly we define an eigenvector of a matrix $\boldsymbol{T}$ to be a non-zero vector $\boldsymbol{v}$ such that $\boldsymbol{T}\boldsymbol{v} = \lambda \boldsymbol{v}$ for some scalar λ; λ is the corresponding eigenvalue of the matrix.

6.1 The characteristic equation

Let λ be an eigenvalue of the linear transformation $T \in L(U, V)$, where V is a vector space of dimension n over the field F. Then there exists a vector $\boldsymbol{v} \neq \boldsymbol{0}$ such that

$$T(\boldsymbol{v}) = \lambda \boldsymbol{v} = \lambda 1 \boldsymbol{v},$$

where 1 is the identity transformation. We may write this equation

$$(\lambda 1 - T)(\boldsymbol{v}) = \boldsymbol{0},$$

and it follows that $\boldsymbol{v} \in \ker (\lambda 1 - T)$. Now $\ker (\lambda 1 - T)$ contains a non-zero vector if and only if $\lambda 1 - T$ is singular (see Theorem 5.1, Corollary 2).

Take a basis for V, and let $\boldsymbol{T}$ be the matrix of T relative to this basis. The matrix of 1 is clearly the unit matrix $\boldsymbol{I}$ so that $\lambda 1 - T$ has matrix $\lambda \boldsymbol{I} - \boldsymbol{T}$. This is singular if and only if $\det (\lambda \boldsymbol{I} - \boldsymbol{T}) = 0$. This gives

$$\begin{vmatrix} \lambda - t_{11} & -t_{12} & \cdots & -t_{1n} \\ -t_{21} & \lambda - t_{22} & \cdots & -t_{2n} \\ \cdot & \cdot & \cdots & \cdot \\ -t_{n1} & -t_{n2} & \cdots & \lambda - t_{nn} \end{vmatrix} = 0.$$

This is clearly a polynomial equation of degree n in λ, of the form

$$\lambda^n + a_1\lambda^{n-1} + a_2\lambda^{n-2} + \cdots + a_{n-1}\lambda + a_n = 0,$$

with coefficients a_i in the field F. The term in λ^{n-1} can come only from the product of the leading diagonal elements

$$(\lambda - t_{11})(\lambda - t_{22}) \cdots (\lambda - t_{nn})$$

and by inspection we see that

$$a_1 = -(t_{11} + t_{22} + \cdots + t_{nn}).$$

The sum of the elements in the leading diagonal of a square matrix $\boldsymbol{T}$ is called the *trace* of $\boldsymbol{T}$, written $\operatorname{tr} \boldsymbol{T}$. Hence $a_1 = -\operatorname{tr} \boldsymbol{T}$.

The term independent of λ is obtained by putting $\lambda = 0$ in the polynomial, so that

$$a_n = \begin{vmatrix} -t_{11} & -t_{12} & \cdots & -t_{1n} \\ -t_{21} & -t_{22} & \cdots & -t_{2n} \\ \cdot & \cdot & \cdots & \cdot \\ -t_{n1} & -t_{n2} & \cdots & -t_{nn} \end{vmatrix} = (-1)^n \det \boldsymbol{T}.$$

If we take a different basis for V, and T has matrix $\boldsymbol{S}$ relative to the new basis, then $\boldsymbol{S} = \boldsymbol{H}^{-1}\boldsymbol{T}\boldsymbol{H}$ (Theorem 3.1), where $\boldsymbol{H}$ is a non-singular $n \times n$ matrix. $\lambda 1 - T$ has matrix $\lambda\boldsymbol{I} - \boldsymbol{S}$ relative to the new basis and

$$\begin{aligned}\det(\lambda\boldsymbol{I} - \boldsymbol{S}) &= \det(\lambda\boldsymbol{I} - \boldsymbol{H}^{-1}\boldsymbol{T}\boldsymbol{H}) \\ &= \det(\lambda\boldsymbol{H}^{-1}\boldsymbol{I}\boldsymbol{H} - \boldsymbol{H}^{-1}\boldsymbol{T}\boldsymbol{H}) \\ &= \det[\boldsymbol{H}^{-1}(\lambda\boldsymbol{I} - \boldsymbol{T})\boldsymbol{H}] \\ &= \det\boldsymbol{H}^{-1} . \det(\lambda\boldsymbol{I} - \boldsymbol{T}) . \det\boldsymbol{H} \\ &= \det(\lambda\boldsymbol{I} - \boldsymbol{T}),\end{aligned}$$

since $\det\boldsymbol{H}^{-1} . \det\boldsymbol{H} = 1$.

It follows that, whatever basis we take for V, we always arrive at the same equation for λ:

$$\lambda^n + a_1\lambda^{n-1} + a_2\lambda^{n-2} + \cdots + a_{n-1}\lambda + a_n = 0. \tag{1}$$

It is called the *characteristic equation* of the transformation T, and the polynomial on the left-hand side of Eqn. (1) is the *characteristic polynomial* of T.

We have therefore shown that the eigenvalues of T are the roots of its characteristic equation. We now see the importance of the field F. Suppose the transformation T has characteristic equation

$$\lambda^5 - \lambda^4 - 4\lambda^3 + 4\lambda^2 - 5\lambda + 5 = 0,$$

i.e.,

$$(\lambda - 1)(\lambda^2 - 5)(\lambda^2 + 1) = 0.$$

If the field is $\boldsymbol{Q}$, the field of rational numbers, the equation has only one root, $\lambda = 1$, in $\boldsymbol{Q}$, and this is the only eigenvalue of T. If the field is $\boldsymbol{R}$, the real numbers, T has three eigenvalues $1, \sqrt{5}, -\sqrt{5}$. If the field is $\boldsymbol{C}$, the complex numbers, T has five eigenvalues $1, \sqrt{5}, -\sqrt{5}, i, -i$.

When the field is $\boldsymbol{C}$, the characteristic equation, Eqn. (1), has exactly n roots in $\boldsymbol{C}$, provided that we count multiple roots according to their multiplicity. If these roots are $\lambda_1, \lambda_2, \ldots, \lambda_n$, not necessarily all distinct, then the characteristic equation is

$$(\lambda - \lambda_1)(\lambda - \lambda_2)\ldots(\lambda - \lambda_n) = 0,$$

or $$\lambda^n - (\lambda_1 + \lambda_2 + \cdots + \lambda_n)\lambda^{n-1} + \cdots + (-1)^n\lambda_1\lambda_2\ldots\lambda_n = 0.$$

Identifying this with Eqn. (1) we have

$$\begin{aligned} a_1 &= -(\lambda_1 + \lambda_2 \cdots + \lambda_n), \\ a_n &= (-1)^n\lambda_1\lambda_2\ldots\lambda_n.\end{aligned}$$

But $a_1 = -\text{tr } \boldsymbol{T}$, $a_n = (-1)^n \det \boldsymbol{T}$ and hence

$$\text{tr } \boldsymbol{T} = \lambda_1 + \lambda_2 + \cdots + \lambda_n = \text{the sum of the eigenvalues of } T,$$
$$\det \boldsymbol{T} = \lambda_1\lambda_2 \ldots \lambda_n = \text{the product of the eigenvalues of } T.$$

If μ is a root of the characteristic equation of multiplicity p, we say that the eigenvalue μ has *multiplicity* p.

6.2 Eigenspaces

Let λ be an eigenvalue of the linear transformation T on V. The set of all eigenvectors $\boldsymbol{x}$ corresponding to this eigenvalue, together with the zero vector, forms a subspace of V called the *eigenspace* corresponding to λ. The proof of this is immediate, since $T(\boldsymbol{x}_1) = \lambda\boldsymbol{x}_1$ and $T(\boldsymbol{x}_2) = \lambda\boldsymbol{x}_2$ implies $T(\boldsymbol{x}_1 + \boldsymbol{x}_2) = \lambda(\boldsymbol{x}_1 + \boldsymbol{x}_2)$ and $T(\boldsymbol{x}) = \lambda\boldsymbol{x}$ implies $T(c\boldsymbol{x}) = \lambda(c\boldsymbol{x})$ for every scalar c.

The eigenspace of T corresponding to the eigenvalue μ is clearly the kernel of the mapping $(\mu 1 - T)$, so that the dimension of the eigenspace is the nullity of the mapping $\mu 1 - T$. Let the eigenspace be W, a subspace of V, and let its dimension be k. Let the eigenvalue μ have multiplicity p. Let $\{\boldsymbol{w}_1, \boldsymbol{w}_2, \ldots, \boldsymbol{w}_k\}$ be a basis for W, and extend this to a basis $\{\boldsymbol{w}_1, \boldsymbol{w}_2, \ldots, \boldsymbol{w}_k, \boldsymbol{w}_{k+1}, \ldots, \boldsymbol{w}_n\}$ for V. Then

$$T(\boldsymbol{w}_i) = \mu\boldsymbol{w}_i \qquad (i = 1, 2, \ldots, k),$$

so that the matrix of T relative to this basis is

$$\begin{bmatrix} \mu & 0 & \cdots 0 & \cdots 0 \\ 0 & \mu & \cdots 0 & \cdots 0 \\ \cdot & \cdot & \cdots\cdot & \cdots\cdot \\ 0 & 0 & \cdots \mu & \cdots 0 \\ t_{k+1,1} & t_{k+1,2} & \cdots t_{k+1,k} & \cdots t_{k+1,n} \\ \cdot & \cdot & \cdots\cdot & \cdots\cdot \\ t_{n,1} & t_{n,2} & \cdots t_{n,k} & \cdots t_{n,n} \end{bmatrix}.$$

The leading submatrix of order $k \times k$ is $\mu\boldsymbol{I}_k$, and we can write the matrix $\boldsymbol{T}$ in partitioned form

$$\boldsymbol{T} = \begin{bmatrix} \mu\boldsymbol{I}_k & \boldsymbol{0} \\ \boldsymbol{A} & \boldsymbol{B} \end{bmatrix},$$

where $\boldsymbol{A}$ is of order $(n-k) \times k$, $\boldsymbol{B}$ is of order $(n-k) \times (n-k)$. Now

$$\det(\lambda \boldsymbol{I} - \boldsymbol{T}) = \begin{vmatrix} \lambda \boldsymbol{I}_k - \mu \boldsymbol{I}_k & \boldsymbol{0} \\ \boldsymbol{A} & \lambda \boldsymbol{I}_{n-k} - \boldsymbol{B} \end{vmatrix}$$
$$= (\lambda - \mu)^k \det(\lambda \boldsymbol{I}_{n-k} - \boldsymbol{B}),$$

so that $(\lambda - \mu)^k$ is a factor of the characteristic polynomial of T. It follows that the multiplicity of the eigenvalue μ is at least k, or $p \geqslant k$. We have therefore proved the following theorem.

THEOREM 6.1. The dimension of an eigenspace cannot exceed the multiplicity of the corresponding eigenvalue.

We now consider the relationship between eigenvectors corresponding to different eigenvalues.

THEOREM 6.2. Eigenvectors corresponding to different eigenvalues are linearly independent.

Proof. The result is proved by induction. Let $\lambda_1, \lambda_2, \ldots, \lambda_k$ be distinct eigenvalues of the linear transformation T on V, and let $\boldsymbol{v}_1, \boldsymbol{v}_2, \ldots, \boldsymbol{v}_k$ be eigenvectors corresponding to $\lambda_1, \lambda_2, \ldots, \lambda_k$ respectively. Thus

$$T(\boldsymbol{v}_i) = \lambda_i \boldsymbol{v}_i \qquad (i = 1, 2, \ldots, k).$$

Now $\boldsymbol{v}_1 \neq \boldsymbol{0}$, so that $\boldsymbol{v}_1$ is certainly linearly independent. Suppose that $\boldsymbol{v}_1, \boldsymbol{v}_2, \ldots, \boldsymbol{v}_m$ are linearly independent $(1 \leqslant m < k)$ and that

$$c_1\boldsymbol{v}_1 + \cdots + c_m\boldsymbol{v}_m + c_{m+1}\boldsymbol{v}_{m+1} = \boldsymbol{0} \qquad (c_1, \ldots, c_{m+1} \in F). \quad (2)$$

Then $$T(c_1\boldsymbol{v}_1 + \cdots + c_{m+1}\boldsymbol{v}_{m+1}) = T(\boldsymbol{0}) = \boldsymbol{0},$$

i.e., $$c_1T(\boldsymbol{v}_1) + \cdots + c_mT(\boldsymbol{v}_m) + c_{m+1}T(\boldsymbol{v}_{m+1}) = \boldsymbol{0},$$

so that $$c_1\lambda_1\boldsymbol{v}_1 + \cdots + c_m\lambda_m\boldsymbol{v}_m + c_{m+1}\lambda_{m+1}\boldsymbol{v}_{m+1} = \boldsymbol{0}. \quad (3)$$

If $\lambda_{m+1} = 0$, in which case $\lambda_1, \ldots, \lambda_m$ are all non-zero, Eqn. (3) becomes

$$c_1\lambda_1\boldsymbol{v}_1 + \cdots + c_m\lambda_m\boldsymbol{v}_m = \boldsymbol{0}.$$

Since $\boldsymbol{v}_1, \boldsymbol{v}_2, \ldots, \boldsymbol{v}_m$ are linearly independent, by hypothesis, this implies

$$c_1\lambda_1 = c_2\lambda_2 = \cdots = c_m\lambda_m = 0,$$

so that $$c_1 = c_2 = \cdots = c_m = 0.$$

Equation (2) then gives $c_{m+1}\boldsymbol{v}_{m+1} = \boldsymbol{0}$, and since $\boldsymbol{v}_{m+1} \neq \boldsymbol{0}$, $c_{m+1} = 0$. Hence Eqn. (2) implies that $c_1 = c_2 = \cdots = c_m = c_{m+1} = 0$, so that $\boldsymbol{v}_1, \ldots, \boldsymbol{v}_m, \boldsymbol{v}_{m+1}$ are linearly independent.

If $\lambda_{m+1} \neq 0$, subtracting λ_{m+1} times Eqn. (2) from Eqn. (3) gives

$$c_1(\lambda_1 - \lambda_{m+1})\boldsymbol{v}_1 + c_2(\lambda_2 - \lambda_{m+1})\boldsymbol{v}_2 + \cdots + c_m(\lambda_m - \lambda_{m+1})\boldsymbol{v}_m = \boldsymbol{0},$$

and, since $\boldsymbol{v}_1, \boldsymbol{v}_2, \ldots, \boldsymbol{v}_m$ are linearly independent, this implies that

$$c_1(\lambda_1 - \lambda_{m+1}) = c_2(\lambda_2 - \lambda_{m+1}) = \cdots = c_m(\lambda_m - \lambda_{m+1}) = 0.$$

But the eigenvalues are distinct and hence

$$c_1 = c_2 = \cdots = c_m = 0.$$

As before this implies that $c_{m+1} = 0$, and $\boldsymbol{v}_1, \ldots, \boldsymbol{v}_m, \boldsymbol{v}_{m+1}$ are linearly independent.

We have proved that if $\boldsymbol{v}_1, \boldsymbol{v}_2, \ldots, \boldsymbol{v}_m$ are linearly independent, then so are $\boldsymbol{v}_1, \boldsymbol{v}_2, \ldots, \boldsymbol{v}_m, \boldsymbol{v}_{m+1}$. But $\boldsymbol{v}_1$ is linearly independent. Hence $\boldsymbol{v}_1, \boldsymbol{v}_2$; $\boldsymbol{v}_1, \boldsymbol{v}_2, \boldsymbol{v}_3$; . . . ; $\boldsymbol{v}_1, \boldsymbol{v}_2, \ldots, \boldsymbol{v}_k$ are linearly independent sets and the result is proved.

6.3 Similar matrices

Two matrices are *similar* if they are the matrices of the *same* linear transformation on a vector space relative to *different* bases. We have proved (Theorem 3.1) that if the linear transformation T on an n-dimensional vector space V has matrices $\boldsymbol{T}$, $\boldsymbol{S}$ relative to different bases, then

$$\boldsymbol{S} = \boldsymbol{H}^{-1}\boldsymbol{T}\boldsymbol{H},$$

where $\boldsymbol{H}$ is a non-singular $n \times n$ matrix. Conversely, if T has matrix $\boldsymbol{T}$ relative to the basis $\{\boldsymbol{v}_1, \boldsymbol{v}_2, \ldots, \boldsymbol{v}_n\}$, then $\boldsymbol{S} = \boldsymbol{H}^{-1}\boldsymbol{T}\boldsymbol{H}$ is the matrix of T relative to the basis $\{\boldsymbol{w}_1, \boldsymbol{w}_2, \ldots, \boldsymbol{w}_n\}$, where $\boldsymbol{w} = \boldsymbol{v}\boldsymbol{H}$.

It follows that two $n \times n$ matrices $\boldsymbol{S}$ and $\boldsymbol{T}$ are similar if and only if $\boldsymbol{S} = \boldsymbol{H}^{-1}\boldsymbol{T}\boldsymbol{H}$ for some non-singular matrix $\boldsymbol{H}$. We can now determine when it is possible to choose a basis for V relative to which $T \in L(V, V)$ has a diagonal matrix.

THEOREM 6.3. An $n \times n$ matrix $\boldsymbol{T}$ is similar to a diagonal matrix if and only if its eigenvectors span the space V_n.

Proof. Relative to the basis $\{\boldsymbol{e}_1, \boldsymbol{e}_2, \ldots, \boldsymbol{e}_n\}$ consisting of the fundamental unit vectors $\boldsymbol{T}$ is the matrix of a unique linear transformation T. First, suppose that the eigenvectors of $\boldsymbol{T}$ span the space V_n. Then we can choose a basis $\{\boldsymbol{x}_1, \boldsymbol{x}_2, \ldots, \boldsymbol{x}_n\}$ of V_n consisting of eigenvectors of $\boldsymbol{T}$. It then follows that there exist scalars $\lambda_1, \lambda_2, \ldots, \lambda_n$ such that

$$\boldsymbol{T}\boldsymbol{x}_i = \lambda_i \boldsymbol{x}_i \qquad (i = 1, 2, \ldots, n). \tag{4}$$

Let $\boldsymbol{x}_1, \boldsymbol{x}_2, \ldots, \boldsymbol{x}_n$ be the coordinate vectors in V_n of the vectors $\boldsymbol{v}_1, \boldsymbol{v}_2, \ldots, \boldsymbol{v}_n$ in the n-dimensional vector space V, relative to the basis $\{\boldsymbol{e}_1, \boldsymbol{e}_2, \ldots, \boldsymbol{e}_n\}$. Then Eqn. (4) can be written

$$T(\boldsymbol{v}_i) = \lambda_i \boldsymbol{v}_i \qquad (i = 1, 2, \ldots, n)$$

and the matrix of T relative to the basis $\{\boldsymbol{v}_1, \boldsymbol{v}_2, \ldots, \boldsymbol{v}_n\}$ of V consisting of eigenvectors of T is the diagonal matrix

$$\Lambda = \begin{bmatrix} \lambda_1 & 0 \cdots 0 \\ 0 & \lambda_2 \cdots 0 \\ \cdot & \cdots\cdots \\ 0 & 0 \cdots \lambda_n \end{bmatrix}.$$

Hence $\boldsymbol{T}$ is similar to Λ.

If $\boldsymbol{X}$ is the $n \times n$ matrix having $\boldsymbol{x}_1, \boldsymbol{x}_2, \ldots, \boldsymbol{x}_n$ as its columns, Eqn. (4) can be written

$$\boldsymbol{T}\boldsymbol{X} = \boldsymbol{X}\Lambda.$$

$\boldsymbol{X}$ is non-singular (since its columns are linearly independent) and $\boldsymbol{X}^{-1}$ exists; hence

$$\boldsymbol{X}^{-1}\boldsymbol{T}\boldsymbol{X} = \boldsymbol{X}^{-1}\boldsymbol{X}\Lambda = \Lambda.$$

Λ is thus of the form $\boldsymbol{H}^{-1}\boldsymbol{T}\boldsymbol{H}$, with $\boldsymbol{H} = \boldsymbol{X}$.

Conversely, suppose that $\boldsymbol{T}$ is similar to the diagonal matrix Λ. Then

$$\boldsymbol{X}^{-1}\boldsymbol{T}\boldsymbol{X} = \Lambda$$

for some non-singular matrix $\boldsymbol{X}$, or

$$\boldsymbol{T}\boldsymbol{X} = \boldsymbol{X}\Lambda.$$

This is equivalent to

$$\boldsymbol{T}\boldsymbol{x}_i = \lambda_i \boldsymbol{x}_i \qquad (i = 1, 2, \ldots, n),$$

where $\boldsymbol{x}_i$, which is clearly an eigenvector of $\boldsymbol{T}$, is the ith column of $\boldsymbol{X}$.

Now $\boldsymbol{x}_1, \boldsymbol{x}_2, \ldots, \boldsymbol{x}_n$ are linearly independent (being the columns of a non-singular matrix) and hence span the space V_n. Thus the eigenvectors of $\boldsymbol{T}$ span the space V_n.

The above proof shows that $X^{-1}TX = \Lambda$, a diagonal matrix, if and only if the columns of X are linearly independent eigenvectors of T, and the diagonal elements of Λ are the eigenvalues of T.

Corollary. Two diagonal matrices are similar if and only if they differ only in the order of their diagonal terms. This follows at once, since their diagonal elements must be the eigenvalues of the corresponding linear transformation.

From Theorems 6.1 and 6.2 we can at once deduce that any $n \times n$ matrix with n distinct eigenvalues is similar to a diagonal matrix. For the matrix has n linearly independent eigenvectors which form a basis for the space V_n.

We now prove a well-known result called the Cayley–Hamilton theorem. This states that a matrix satisfies its characteristic equation and is to be interpreted in the following sense. Let the $n \times n$ matrix A have characteristic equation

$$\lambda_n + a_1\lambda^{n-1} + a_2\lambda^{n-2} + \cdots + a_{n-1}\lambda + a_n = 0.$$

Then

$$A^n + a_1A^{n-1} + a_2A^{n-2} + \cdots a_{n-1}A + a_nI = 0,$$

where the zero on the right-hand side is now the $n \times n$ zero matrix.

THEOREM 6.4. (Cayley–Hamilton) A matrix satisfies its characteristic equation.

Proof. Let A be an $n \times n$ matrix with elements in the field F and with characteristic equation

$$\lambda^n + a_1\lambda^{n-1} + \cdots + a_{n-1}\lambda + a_n = 0.$$

Let $B = \text{adj}\,(\lambda I - A)$, where λ is any scalar in F. The elements of the matrix B are all polynomials in λ of degree $\leqslant n - 1$. Hence we can write

$$B = B_0 + B_1\lambda + B_2\lambda^2 + \cdots + B_{n-1}\lambda^{n-1},$$

where $B_0, B_1, \ldots, B_{n-1}$ are all $n \times n$ matrices with elements in F. Now

$$(\lambda I - A)\,.\,\text{adj}\,(\lambda I - A) = \det\,(\lambda I - A)\,.\,I;$$

i.e.,

$$(\lambda I - A)(B_0 + B_1\lambda + \cdots + B_{n-1}\lambda^{n-1}) = (\lambda^n + a_1\lambda^{n-1} + \cdots + a_n)I.$$

This is true for all scalars $\lambda \in F$. Hence, equating coefficients of like powers of λ,

$$\begin{aligned} \boldsymbol{B}_{n-1} &= \boldsymbol{I}, \\ \boldsymbol{B}_{n-2} - \boldsymbol{A}\boldsymbol{B}_{n-1} &= a_1\boldsymbol{I}, \\ \boldsymbol{B}_{n-3} - \boldsymbol{A}\boldsymbol{B}_{n-2} &= a_2\boldsymbol{I}, \\ &\cdots\cdots \\ \boldsymbol{B}_0 - \boldsymbol{A}\boldsymbol{B}_1 &= a_{n-1}\boldsymbol{I}, \\ -\boldsymbol{A}\boldsymbol{B}_0 &= a_n\boldsymbol{I}. \end{aligned}$$

Multiplying these equations by $\boldsymbol{A}^n, \boldsymbol{A}^{n-1}, \ldots, \boldsymbol{A}, \boldsymbol{I}$ respectively and adding gives

$$\begin{aligned} &\boldsymbol{A}^n + a_1\boldsymbol{A}^{n-1} + a_2\boldsymbol{A}^{n-2} + \cdots + a_{n-1}\boldsymbol{A} + a_n\boldsymbol{I} \\ &= \boldsymbol{A}^n\boldsymbol{B}_{n-1} + \boldsymbol{A}^{n-1}(\boldsymbol{B}_{n-2} - \boldsymbol{A}\boldsymbol{B}_{n-1}) + \boldsymbol{A}^{n-2}(\boldsymbol{B}_{n-3} - \boldsymbol{A}\boldsymbol{B}_{n-2}) \\ &\qquad + \cdots + \boldsymbol{A}(\boldsymbol{B}_0 - \boldsymbol{A}\boldsymbol{B}_1) - \boldsymbol{A}\boldsymbol{B}_0 \\ &= \boldsymbol{0}. \end{aligned}$$

We can use Theorem 6.4 to express $\boldsymbol{A}^k$, $k \geqslant n$, as a polynomial in $\boldsymbol{A}$ of degree at most $n - 1$. For

$$\begin{aligned} \boldsymbol{A}^{k-n}(\boldsymbol{A}^n + a_1\boldsymbol{A}^{n-1} + \cdots + a_n\boldsymbol{I}) &= \boldsymbol{A}^k + a_1\boldsymbol{A}^{k-1} + \cdots + a_n\boldsymbol{A}^{k-n} \\ &= \boldsymbol{0}, \end{aligned}$$

and we can express $\boldsymbol{A}^k$ as a polynomial in $\boldsymbol{A}$ of degree at most $k - 1$. Repeating this process with the term in $\boldsymbol{A}^{k-1}$ we obtain a polynomial of degree at most $k - 2$, and so on.

Example 1. Where possible, find diagonal matrices similar to the matrices

$$\boldsymbol{A} = \begin{bmatrix} -1 & 6 & -12 \\ 0 & -13 & 30 \\ 0 & -9 & 20 \end{bmatrix}, \qquad \boldsymbol{B} = \begin{bmatrix} 2 & -1 & 0 \\ 1 & 0 & 0 \\ 0 & 0 & 3 \end{bmatrix},$$

$$\boldsymbol{C} = \begin{bmatrix} 17 & -10 & -5 \\ 45 & -28 & -15 \\ -30 & 20 & 12 \end{bmatrix}, \qquad \boldsymbol{D} = \begin{bmatrix} 1 & -5 \\ 1 & -1 \end{bmatrix}.$$

The characteristic equation of the matrix $\boldsymbol{A}$ is

$$\begin{bmatrix} -1-\lambda & 6 & -12 \\ 0 & -13-\lambda & 30 \\ 0 & -9 & 20-\lambda \end{bmatrix} = 0,$$

i.e.,

$$(\lambda + 1)(\lambda - 2)(\lambda - 5) = 0.$$

The eigenvalues are -1, 2, 5. Since these are distinct, the corresponding eigenvectors are linearly independent and therefore span the space $\boldsymbol{R}_3$. Hence $\boldsymbol{A}$ is similar to the diagonal matrix

$$\begin{bmatrix} -1 & 0 & 0 \\ 0 & 2 & 0 \\ 0 & 0 & 5 \end{bmatrix} = \Lambda.$$

The reader should check that $\{1\ 0\ 0\}$, $\{0\ 2\ 1\}$, $\{-1\ 5\ 3\}$ are eigenvectors corresponding to the eigenvalues -1, 2, 5 respectively. Taking

$$\boldsymbol{H} = \begin{bmatrix} 1 & 0 & -1 \\ 0 & 2 & 5 \\ 0 & 1 & 3 \end{bmatrix},$$

$$\boldsymbol{H}^{-1}\boldsymbol{A}\boldsymbol{H} = \Lambda.$$

The characteristic equation of $\boldsymbol{B}$ is

$$\begin{vmatrix} 2-\lambda & -1 & 0 \\ 1 & -\lambda & 0 \\ 0 & 0 & 3-\lambda \end{vmatrix} = 0,$$

i.e.,
$$(\lambda - 3)(\lambda - 1)^2 = 0.$$

The eigenvalues are 3, 1, 1.

When $\lambda = 1$, the matrix $\boldsymbol{B} - \lambda\boldsymbol{I} = \begin{bmatrix} 1 & -1 & 0 \\ 1 & -1 & 0 \\ 0 & 0 & 2 \end{bmatrix}$ has rank 2. Hence the equations $(\boldsymbol{B} - \lambda\boldsymbol{I})\boldsymbol{x} = \boldsymbol{0}$ have only $3 - 2 = 1$ linearly independent solution. Thus the eigenvectors of $\boldsymbol{B}$ do *not* span the space $\boldsymbol{R}_3$ and we cannot find a diagonal matrix similar to $\boldsymbol{B}$.

The characteristic equation of $\boldsymbol{C}$ is

$$\begin{vmatrix} 17-\lambda & -10 & -5 \\ 45 & -28-\lambda & -15 \\ -30 & 20 & 12-\lambda \end{vmatrix} = 0.$$

Subtracting three times the first row from the second and adding twice the first row to the last row this becomes

$$\begin{vmatrix} 17-\lambda & -10 & -5 \\ -6+3\lambda & 2-\lambda & 0 \\ 4-2\lambda & 0 & 2-\lambda \end{vmatrix} = 0,$$

i.e.,
$$(2-\lambda)^2 \begin{vmatrix} 17-\lambda & -10 & -5 \\ -3 & 1 & 0 \\ 2 & 0 & 1 \end{vmatrix} = 0,$$

i.e.,
$$(2-\lambda)^2 (3+\lambda) = 0.$$

Thus the eigenvalues of $\boldsymbol{C}$ are $-3, 2, 2$.

When $\lambda = 2$, $\boldsymbol{C} - \lambda\boldsymbol{I} = \begin{bmatrix} 15 & -10 & -5 \\ 45 & -30 & -15 \\ -30 & 20 & 10 \end{bmatrix}$ which has rank 1, since each row is a multiple of $[3\ \ -2\ \ -1]$. It follows that $(\boldsymbol{C} - \lambda\boldsymbol{I})\boldsymbol{x} = \boldsymbol{0}$ has $3 - 1 = 2$ linearly independent solutions in $\boldsymbol{R}_3$ and the eigenspace corresponding to $\lambda = 2$ is of dimension 2. Thus the eigenvectors of $\boldsymbol{C}$ span the space $\boldsymbol{R}_3$, and $\boldsymbol{C}$ is therefore similar to the diagonal matrix $\begin{bmatrix} -3 & 0 & 0 \\ 0 & 2 & 0 \\ 0 & 0 & 2 \end{bmatrix} = \Lambda$. The vectors $\{1\ \ 3\ \ -2\}$, $\{1\ \ 0\ \ 3\}$, $\{2\ \ 3\ \ 0\}$ are eigenvectors corresponding to the eigenvalues $-3, 2, 2$ respectively, and we may, for example, take

$$\boldsymbol{H} = \begin{bmatrix} 1 & 1 & 2 \\ 3 & 0 & 3 \\ -2 & 3 & 0 \end{bmatrix}.$$

Then
$$\boldsymbol{H}^{-1}\boldsymbol{C}\boldsymbol{H} = \Lambda.$$

The characteristic equation of $\boldsymbol{D}$ is

$$\begin{vmatrix} 1-\lambda & -5 \\ 1 & -1-\lambda \end{vmatrix} = 0,$$

i.e.,
$$\lambda^2 + 4 = 0.$$

If the underlying field is $\boldsymbol{R}$, the matrix has no eigenvalues and no eigenvectors, and there is no diagonal matrix similar to $\boldsymbol{D}$, having its elements in $\boldsymbol{R}$. If, however, the underlying field is $\boldsymbol{C}$, the matrix has two distinct eigenvalues $+2i$ and $-2i$, and it is therefore similar to the diagonal matrix $\begin{bmatrix} 2i & 0 \\ 0 & -2i \end{bmatrix}$. The reader should find the corresponding matrix $\boldsymbol{H}$ in this case.

Example 2. Find $\boldsymbol{A}^7$, when $\boldsymbol{A} = \begin{bmatrix} -4 & 6 \\ -3 & 5 \end{bmatrix}$.

The characteristic equation of $\boldsymbol{A}$ is

$$(-4 - \lambda)(5 - \lambda) + 18 = 0,$$

i.e.,

$$\lambda^2 - \lambda - 2 = 0.$$

Thus, by the Cayley–Hamilton theorem,

$$\boldsymbol{A}^2 - \boldsymbol{A} - 2\boldsymbol{I} = \boldsymbol{0}.$$

Hence

$$\begin{aligned} \boldsymbol{A}^7 &= \boldsymbol{A}^6 + 2\boldsymbol{A}^5 = 3\boldsymbol{A}^5 + 2\boldsymbol{A}^4 \\ &= 5\boldsymbol{A}^4 + 6\boldsymbol{A}^3 = 11\boldsymbol{A}^3 + 10\boldsymbol{A}^2 \\ &= 21\boldsymbol{A}^2 + 22\boldsymbol{A} = 43\boldsymbol{A} + 42\boldsymbol{I} \\ &= \begin{bmatrix} -130 & 258 \\ -129 & 257 \end{bmatrix}. \end{aligned}$$

Alternatively we may proceed as follows. $\{2\ 1\}$, $\{1\ 1\}$ are eigenvectors corresponding to the eigenvalues -1, 2 respectively. Taking

$$\boldsymbol{H} = \begin{bmatrix} 2 & 1 \\ 1 & 1 \end{bmatrix}, \qquad \boldsymbol{H}^{-1} = \begin{bmatrix} 1 & -1 \\ -1 & 2 \end{bmatrix},$$

$$\boldsymbol{H}^{-1}\boldsymbol{A}\boldsymbol{H} = \begin{bmatrix} -1 & 0 \\ 0 & 2 \end{bmatrix},$$

$$(\boldsymbol{H}^{-1}\boldsymbol{A}\boldsymbol{H})^7 = \boldsymbol{H}^{-1}\boldsymbol{A}^7\boldsymbol{H} = \begin{bmatrix} -1 & 0 \\ 0 & 2^7 \end{bmatrix},$$

$$\begin{aligned} \boldsymbol{A}^7 &= \begin{bmatrix} 2 & 1 \\ 1 & 1 \end{bmatrix} \begin{bmatrix} -1 & 0 \\ 0 & 128 \end{bmatrix} \begin{bmatrix} 1 & -1 \\ -1 & 2 \end{bmatrix} \\ &= \begin{bmatrix} -130 & 258 \\ -129 & 257 \end{bmatrix}. \end{aligned}$$

Example 3. Prove that the matrices

$$\boldsymbol{A} = \begin{bmatrix} -10 & 6 & 3 \\ -26 & 16 & 8 \\ 16 & -10 & -5 \end{bmatrix} \quad \text{and} \quad \boldsymbol{B} = \begin{bmatrix} 0 & -6 & -16 \\ 0 & 17 & 45 \\ 0 & -6 & -16 \end{bmatrix}$$

are similar.

A and B both have the three distinct eigenvalues 0, 2, -1, and each is therefore similar to the diagonal matrix $\Lambda = \begin{bmatrix} 0 & 0 & 0 \\ 0 & 2 & 0 \\ 0 & 0 & -1 \end{bmatrix}$. Hence there exist non-singular matrices H, K such that

$$H^{-1}AH = K^{-1}BK = \Lambda.$$

Thus

$$B = KH^{-1}AHK^{-1}.$$

Writing $M = HK^{-1}$, so that $M^{-1} = KH^{-1}$, we have

$$B = M^{-1}AM,$$

and B is similar to A.

Problems

6.1 Find the eigenvalues and eigenvectors of the matrices

$$A = \begin{bmatrix} 3 & 4 \\ -2 & -3 \end{bmatrix}, \quad B = \begin{bmatrix} 2 & \sqrt{2} \\ \sqrt{2} & 1 \end{bmatrix}, \quad C = \begin{bmatrix} 4 & -2 \\ 2 & 0 \end{bmatrix}.$$

The underlying field is R.

6.2 Determine the dimension of each eigenspace of the matrix X and if possible find a diagonal matrix that is similar to X when (*a*) $X = A$, (*b*) $X = B$, (*c*) $X = C$. A, B, C are the matrices of Problem 6.1.

6.3 Repeat Problems 6.1. and 6.2 when

$$A = \begin{bmatrix} -2 & 9 & -6 \\ 1 & -2 & 0 \\ 3 & -9 & 5 \end{bmatrix}, \quad B = \begin{bmatrix} 2 & -1 & -1 \\ 0 & 3 & 2 \\ -1 & 1 & 2 \end{bmatrix},$$

$$C = \begin{bmatrix} 1+i & 0 & 0 \\ 2-2i & 1-i & 0 \\ 2i & 0 & 1 \end{bmatrix}.$$

For A and B the underlying field is R and for C it is C.

6.4 Find the characteristic equation of the matrix A of Problem 6.1, and verify that A satisfies this equation. Write down the matrices A^7, A^8.

6.5 Find the characteristic equation of the matrix B of Problem 6.1 and prove that $B^9 = 3^8 \,.\, B$.

6.6 If $A = \begin{bmatrix} -25 & -36 \\ 18 & 26 \end{bmatrix}$, find a matrix H such that $H^{-1}AH$ is diagonal. Hence find A^{12}.

6.7 Prove that the matrices P and Q are similar, where

$$P = \begin{bmatrix} -2 & 0 \\ -14 & 5 \end{bmatrix}, \quad Q = \begin{bmatrix} 12 & -7 \\ 14 & -9 \end{bmatrix}.$$

6.8 A is an $n \times n$ matrix with real elements. Prove that, if λ is an eigenvalue of A, then λ^k is an eigenvalue of A^k, where k is a positive integer. Deduce that $p(\lambda)$ is an eigenvalue of $p(A)$, where p is a polynomial in λ with real coefficients.

Find the eigenvalues of the matrix

$$A = \begin{bmatrix} 1 & 3 \\ 5 & 3 \end{bmatrix}$$

and verify directly that the eigenvalues of A^2 are the squares of the eigenvalues of A.

6.9 A is an $n \times n$ matrix and $A^2 = \boldsymbol{0}$. Use the result of Problem 6.8 to show that A has no non-zero eigenvalues.

7 · *Euclidean and Unitary Spaces*

The set of all points in a plane may be regarded as a vector space of dimension two. Some of the geometrical properties of the plane may be expressed as properties of the vector space. For example, let the points P, Q, R in a plane be represented by the vectors $\boldsymbol{p}$, $\boldsymbol{q}$, $\boldsymbol{r}$ respectively. Then P, Q, R are collinear if and only if the lines PQ, PR have the same direction, i.e., if and only if the vectors $\boldsymbol{q} - \boldsymbol{p}$, $\boldsymbol{r} - \boldsymbol{p}$ are linearly dependent. The same is true of three points in three-dimensional space. Again, the four points P, Q, R, S are coplanar if and only if the lines PQ, PR, and PS are coplanar, i.e., if and only if the vectors $\boldsymbol{q} - \boldsymbol{p}$, $\boldsymbol{r} - \boldsymbol{p}$, $\boldsymbol{s} - \boldsymbol{p}$ are linearly dependent. There are, however, certain geometrical concepts such as angle and distance which we cannot, as yet, interpret in terms of a vector space. We now seek to remedy this omission.

7.1 Distance and norm

Let the real number $d = d(\mathrm{P}, \mathrm{Q})$ be the distance between the points P and Q. We require the distance function d to satisfy the following conditions:

(1) $d(\mathrm{P}, \mathrm{Q}) = d(\mathrm{Q}, \mathrm{P})$;

(2) $d(\mathrm{P}, \mathrm{R}) \leqslant d(\mathrm{P}\ \mathrm{Q}) + d(\mathrm{Q}, \mathrm{R})$;

(3) $d(\mathrm{P}, \mathrm{Q}) > 0$ if $\mathrm{Q} \neq \mathrm{P}$,

$d(\mathrm{P}, \mathrm{P}) = 0$.

(2) is known as the *triangle inequality* since it expresses the fact that the sum of the lengths of two sides of a triangle is never less than the length of the third side. A space in which a distance function satisfying (1), (2), and (3) is defined is called a *metric space*. Even when the space is not geometrical in character in the familiar sense we retain the term distance or *metric*. We give just one example to illustrate this. Let V be a vector space and define a distance function d as follows:

$$d(\boldsymbol{u}, \boldsymbol{v}) = 1, \qquad \boldsymbol{u} \neq \boldsymbol{v}, \qquad \boldsymbol{u}, \boldsymbol{v} \in V,$$
$$d(\boldsymbol{v}, \boldsymbol{v}) = 0 \qquad \text{for all } \boldsymbol{v} \in V.$$

It is easy to see that conditions (1), (2), and (3) are satisfied.

Let V be a vector space over the real field $\boldsymbol{R}$. We wish to define a distance function on V, satisfying (1), (2), and (3). First, consider a mapping of pairs of elements of V into the real numbers $\boldsymbol{R}$. Thus we seek to associate with the pair of elements $\boldsymbol{p}, \boldsymbol{q} \in V$ a real number $(\boldsymbol{p}, \boldsymbol{q})$. Suppose that, for all $\boldsymbol{p}, \boldsymbol{q}, \boldsymbol{r} \in V$, $x \in \boldsymbol{R}$, $(\boldsymbol{p}, \boldsymbol{q})$ has the following properties:

(4) $(\boldsymbol{p}, \boldsymbol{q}) = (\boldsymbol{q}, \boldsymbol{p})$;
(5) $(\boldsymbol{p} + \boldsymbol{q}, \boldsymbol{r}) = (\boldsymbol{p}, \boldsymbol{r}) + (\boldsymbol{q}, \boldsymbol{r})$, $(x\boldsymbol{p}, \boldsymbol{q}) = x(\boldsymbol{p}, \boldsymbol{q})$;
(6) $(\boldsymbol{p}, \boldsymbol{p}) \geqslant 0$ for all $\boldsymbol{p}$,
$(\boldsymbol{p}, \boldsymbol{p}) = 0$ if and only if $\boldsymbol{p} = \boldsymbol{0}$.

(4) states that $(\boldsymbol{p}, \boldsymbol{q})$ is symmetric and (5) that it is linear in the first variable. (5) and (4) together imply that

$$(\boldsymbol{r}, \boldsymbol{p} + \boldsymbol{q}) = (\boldsymbol{r}, \boldsymbol{p}) + (\boldsymbol{r}, \boldsymbol{q}) \qquad \text{and} \qquad (\boldsymbol{q}, x\boldsymbol{p}) = x(\boldsymbol{q}, \boldsymbol{p}).$$

Thus $(\boldsymbol{p}, \boldsymbol{q})$ is linear in the second variable also and we say that it is *bilinear*. (6) states that $(\boldsymbol{p}, \boldsymbol{q})$ is *positive definite*.

A mapping $(\boldsymbol{p}, \boldsymbol{q})$ satisfying conditions (4), (5), and (6) is known as an *inner product* on the space V, and V is called an *inner product space*.

We now define the length or *norm* $\|\boldsymbol{p}\|$ of a vector $\boldsymbol{p}$ by

$$\|\boldsymbol{p}\| = +\sqrt{(\boldsymbol{p}, \boldsymbol{p})}.$$

It follows from (6) that $\|\boldsymbol{p}\|$ is defined and non-negative for all $\boldsymbol{p}$, and $\|\boldsymbol{p}\| = 0$ if and only if $\boldsymbol{p} = \boldsymbol{0}$.

We now show that the function $d(\boldsymbol{p}, \boldsymbol{q}) = \|\boldsymbol{p} - \boldsymbol{q}\|$ is a suitable definition of a distance function. That the function satisfies (3) is immediately obvious from the definition. Now

$$\|-\boldsymbol{p}\|^2 = (-\boldsymbol{p}, -\boldsymbol{p}) = (-1)^2(\boldsymbol{p}, \boldsymbol{p}) = \|\boldsymbol{p}\|^2,$$

so that $\|-\boldsymbol{p}\| = \|\boldsymbol{p}\|$. In particular $\|\boldsymbol{p} - \boldsymbol{q}\| = \|\boldsymbol{q} - \boldsymbol{p}\|$ and condition (1) is satisfied.

Now suppose $\boldsymbol{q} \neq \boldsymbol{0}$, and write $\boldsymbol{r} = \boldsymbol{q}/\|\boldsymbol{q}\|$, so that $\|\boldsymbol{r}\| = 1$. For any vector $\boldsymbol{p}$ we have

$$\begin{aligned} 0 &\leqslant \|\boldsymbol{p} - (\boldsymbol{p}, \boldsymbol{r})\boldsymbol{r}\|^2 = (\boldsymbol{p} - (\boldsymbol{p}, \boldsymbol{r})\boldsymbol{r}, \boldsymbol{p} - (\boldsymbol{p}, \boldsymbol{r})\boldsymbol{r}) \\ &= (\boldsymbol{p}, \boldsymbol{p}) - 2\{(\boldsymbol{p}, \boldsymbol{r})\}^2 + \{(\boldsymbol{p}, \boldsymbol{r})\}^2 \,.\, (\boldsymbol{r}, \boldsymbol{r}) \\ &= \|\boldsymbol{p}\|^2 - \{(\boldsymbol{p}, \boldsymbol{r})\}^2. \end{aligned}$$

Hence

$$\begin{aligned} \{(\boldsymbol{p}, \boldsymbol{r})\}^2 &\leqslant \|\boldsymbol{p}\|^2, \\ |(\boldsymbol{p}, \boldsymbol{r})| &\leqslant \|\boldsymbol{p}\|, \\ |(\boldsymbol{p}, \boldsymbol{q})/\|\boldsymbol{q}\|\,| &\leqslant \|\boldsymbol{p}\|, \\ |(\boldsymbol{p}, \boldsymbol{q})| &\leqslant \|\boldsymbol{p}\| \,.\, \|\boldsymbol{q}\|. \end{aligned} \tag{7}$$

$|(\boldsymbol{p}, \boldsymbol{q})|$ is the modulus, or numerical value, of the real number $(\boldsymbol{p}, \boldsymbol{q})$, in the usual sense. If $\boldsymbol{q} = \boldsymbol{0}$, (7) clearly holds with both sides equal to zero, so that (7) is true for all $\boldsymbol{p}, \boldsymbol{q} \in V$. This inequality is the *Schwarz inequality*. Now

$$\begin{aligned} \|\boldsymbol{p} + \boldsymbol{q}\|^2 &= \|\boldsymbol{p}\|^2 + \|\boldsymbol{q}\|^2 + 2(\boldsymbol{p}, \boldsymbol{q}) \\ &\leqslant \|\boldsymbol{p}\|^2 + \|\boldsymbol{q}\|^2 + 2|(\boldsymbol{p}, \boldsymbol{q})| \\ &\leqslant \|\boldsymbol{p}\|^2 + \|\boldsymbol{q}\|^2 + 2\|\boldsymbol{p}\| \,.\, \|\boldsymbol{q}\| \text{ (by the Schwarz inequality)} \\ &= (\|\boldsymbol{p}\| + \|\boldsymbol{q}\|)^2. \end{aligned}$$

Hence

$$\|\boldsymbol{p} + \boldsymbol{q}\| \leqslant \|\boldsymbol{p}\| + \|\boldsymbol{q}\|.$$

Replacing $\boldsymbol{p}$ by $\boldsymbol{p} - \boldsymbol{q}$ and $\boldsymbol{q}$ by $\boldsymbol{q} - \boldsymbol{r}$ this gives

$$\|\boldsymbol{p} - \boldsymbol{r}\| \leqslant \|\boldsymbol{p} - \boldsymbol{q}\| + \|\boldsymbol{q} - \boldsymbol{r}\|.$$

This is the triangle inequality (2).

7.2 Angle

If $\boldsymbol{p} \neq \boldsymbol{0}$, $\boldsymbol{q} \neq \boldsymbol{0}$, then $\|\boldsymbol{p}\| \neq 0$, $\|\boldsymbol{q}\| \neq 0$ and we can rewrite (7) in the form

$$-1 \leqslant \frac{(\boldsymbol{p}, \boldsymbol{q})}{\|\boldsymbol{p}\|\ \|\boldsymbol{q}\|} \leqslant +1.$$

Now as θ increases steadily from 0 to π, $\cos\theta$ decreases continuously from $+1$ to -1, and hence there is exactly *one* value of θ in the range $0 \leqslant \theta \leqslant \pi$ for which

$$\cos\theta = \frac{(\boldsymbol{p}, \boldsymbol{q})}{\|\boldsymbol{p}\|\ \|\boldsymbol{q}\|}.$$

We define the *angle* between two non-zero vectors $\boldsymbol{p}$ and $\boldsymbol{q}$ to be the unique value of θ in the range $[0, \pi]$ for which

$$\cos \theta = \frac{(\boldsymbol{p}, \boldsymbol{q})}{\|\boldsymbol{p}\| \, \|\boldsymbol{q}\|}.$$

If $\theta = \pi/2$, $\cos \theta = 0$, $(\boldsymbol{p}, \boldsymbol{q}) = 0$, and conversely if $(\boldsymbol{p}, \boldsymbol{q}) = 0$, $\cos \theta = 0$, $\theta = \pi/2$. We therefore say that the vectors $\boldsymbol{p}$ and $\boldsymbol{q}$ are *orthogonal* if and only if $(\boldsymbol{p}, \boldsymbol{q}) = 0$.

7.3 The spaces $\boldsymbol{R}_2$ and $\boldsymbol{R}_3$

We now show that these definitions are consistent with the known results for the spaces $\boldsymbol{R}_2$ and $\boldsymbol{R}_3$.

Let points P, Q in a given plane have rectangular cartesian coordinates (x_1, x_2), (y_1, y_2) respectively, and let $\boldsymbol{p} = \{x_1, x_2\}$, $\boldsymbol{q} = \{y_1, y_2\}$ be their coordinate vectors. Take

$$(\boldsymbol{p}, \boldsymbol{q}) = x_1 y_1 + x_2 y_2.$$

We now show that $(\boldsymbol{p}, \boldsymbol{q})$ is an inner product on the space $\boldsymbol{R}_2$. Clearly $(\boldsymbol{p}, \boldsymbol{q}) = (\boldsymbol{q}, \boldsymbol{p})$.

If R has coordinate vector $\boldsymbol{r} = \{z_1, z_2\}$, then

$$\begin{aligned} (\boldsymbol{p} + \boldsymbol{q}, \boldsymbol{r}) &= (x_1 + y_1)z_1 + (x_2 + y_2)z_2 \\ &= (x_1 z_1 + x_2 z_2) + (y_1 z_1 + y_2 z_2) \\ &= (\boldsymbol{p}, \boldsymbol{r}) + (\boldsymbol{q}, \boldsymbol{r}). \end{aligned}$$

Also, if k is any real number,

$$\begin{aligned} (k\boldsymbol{p}, \boldsymbol{q}) &= (kx_1)y_1 + (kx_2)y_2 \\ &= k(x_1 y_1 + x_2 y_2) \\ &= k(\boldsymbol{p}, \boldsymbol{q}). \end{aligned}$$

Finally, $(\boldsymbol{p}, \boldsymbol{p}) = x_1^2 + x_2^2 \geqslant 0$ for all x_1, x_2. Moreover $x_1^2 + x_2^2 = 0$ if and only if $x_1 = x_2 = 0$, so that $(\boldsymbol{p}, \boldsymbol{p}) = 0$ if and only if $\boldsymbol{p} = \boldsymbol{0}$.

We have shown that $(\boldsymbol{p}, \boldsymbol{q})$ satisfies conditions (4), (5), and (6) and is therefore an inner product on $\boldsymbol{R}_2$. We may now define the length of a vector by

$$\|\boldsymbol{p}\| = +\sqrt{(\boldsymbol{p}, \boldsymbol{p})} = +\sqrt{(x_1^2 + x_2^2)}$$

and the right-hand side is the well-known formula for the length of OP. Again

$$d(\boldsymbol{p}, \boldsymbol{q}) = \|\boldsymbol{p} - \boldsymbol{q}\| = +\sqrt{\{(x_1 - y_1)^2 + (x_2 - y_2)^2\}}$$

and the right-hand side is equal to the length of PQ. Finally

$$\cos\theta = \frac{(\boldsymbol{p}, \boldsymbol{q})}{\|\boldsymbol{p}\|\,\|\boldsymbol{q}\|} = \frac{x_1y_1 + x_2y_2}{\sqrt{(x_1^2 + x_2^2)}\sqrt{(y_1^2 + y_2^2)}}$$

and this is the well-known formula for cos $\angle$ POQ. In particular, OP is perpendicular to OQ if and only if $x_1y_1 + x_2y_2 = 0$.

In $\boldsymbol{R}_3$ let the points P, Q have rectangular cartesian coordinates (x_1, x_2, x_3) and (y_1, y_2, y_3) respectively and define

$$(\boldsymbol{p}, \boldsymbol{q}) = x_1y_1 + x_2y_2 + x_3y_3.$$

Exactly as for $\boldsymbol{R}_2$ we can show that $(\boldsymbol{p}, \boldsymbol{q})$ is an inner product on $\boldsymbol{R}_3$, and we obtain

$$\|\boldsymbol{p}\| = +\sqrt{(x_1^2 + x_2^2 + x_3^2)} = \text{OP},$$

$$\cos\theta = \frac{x_1y_1 + x_2y_2 + x_3y_3}{\sqrt{(x_1^2 + x_2^2 + x_3^2)}\sqrt{(y_1^2 + y_2^2 + y_3^2)}} = \cos\angle\text{POQ}.$$

In particular, OP is perpendicular to OQ if and only if

$$x_1y_1 + x_2y_2 + x_3y_3 = 0.$$

We have therefore shown that our definitions of length, angle, and orthogonality are extensions of (and are consistent with) the corresponding concepts for spaces of two and three dimensions.

7.4 Euclidean spaces

A vector space V over the real field $\boldsymbol{R}$, on which an inner product is defined, is called a *euclidean space.* $\boldsymbol{R}_2$ and $\boldsymbol{R}_3$ with the inner products defined in Section 7.3 are euclidean spaces. The reader should verify that the space $\boldsymbol{R}_n$ with an inner product defined by

$$(\boldsymbol{p}, \boldsymbol{q}) = x_1y_1 + x_2y_2 + \cdots + x_ny_n$$

is a euclidean space.

We now show that an inner product on a vector space of dimension n over $\boldsymbol{R}$ can always be expressed in this form. A vector $\boldsymbol{p}$ is said to be *normal* if it is of unit length, i.e., if $\|\boldsymbol{p}\| = 1$. A set of vectors $\boldsymbol{p}_1, \boldsymbol{p}_2, \ldots, \boldsymbol{p}_k$ is said to form an *orthonormal* (i.e., orthogonal and normal) set if

$$(\boldsymbol{p}_i, \boldsymbol{p}_j) = 0 \qquad (i \neq j),$$
$$\|\boldsymbol{p}_i\| = 1 \qquad (i = 1, 2, \ldots, k).$$

Thus each vector of the set is normal and is orthogonal to every other vector of the set. Now suppose that it is possible to find an orthonormal set of vectors which form a basis for a given vector space V of dimension n. Such a basis is called an *orthonormal basis*. Let $\{\boldsymbol{e}_1, \boldsymbol{e}_2, \ldots, \boldsymbol{e}_n\}$ be an orthonormal basis for V. Thus

$$(\boldsymbol{e}_i, \boldsymbol{e}_j) = 0 \qquad (i \neq j), \qquad \|\boldsymbol{e}_i\| = 1 \qquad (i = 1, 2, \ldots, n).$$

Let $(x_1, x_2, \ldots, x_n)$, $(y_1, y_2, \ldots, y_n)$ be the coordinate vectors of $\boldsymbol{p}$, $\boldsymbol{q}$ relative to this basis, so that

$$\boldsymbol{p} = x_1\boldsymbol{e}_1 + x_2\boldsymbol{e}_2 + \cdots + x_n\boldsymbol{e}_n,$$
$$\boldsymbol{q} = y_1\boldsymbol{e}_1 + y_2\boldsymbol{e}_2 + \cdots + y_n\boldsymbol{e}_n.$$

Using the bilinear property of the scalar product we have

$$\begin{aligned}(\boldsymbol{p}, \boldsymbol{q}) &= (x_1\boldsymbol{e}_1 + \cdots + x_n\boldsymbol{e}_n, y_1\boldsymbol{e}_1 + \cdots + y_n\boldsymbol{e}_n)\\ &= \sum_{k=1}^{n} x_k y_k(\boldsymbol{e}_k, \boldsymbol{e}_k) + \sum_{h \neq k} x_h y_k(\boldsymbol{e}_h, \boldsymbol{e}_k)\\ &= \sum_{k=1}^{n} x_k y_k,\end{aligned}$$

since
$$(\boldsymbol{e}_k, \boldsymbol{e}_k) = \|\boldsymbol{e}_k\|^2 = 1,$$

i.e., $(\boldsymbol{p}, \boldsymbol{q}) = x_1y_1 + x_2y_2 + \cdots + x_ny_n$.

In $\boldsymbol{R}_2$ we took the unit vectors $\boldsymbol{i}, \boldsymbol{j}$ in the directions of the axes as an orthonormal basis and in $\boldsymbol{R}_3$ the vectors $\boldsymbol{i}, \boldsymbol{j}, \boldsymbol{k}$. Later in the chapter we shall show that every euclidean space of finite dimension has an orthonormal basis and we shall also show how to construct such a basis.

7.5 Unitary spaces

We now extend these concepts to vector spaces V over the field $\boldsymbol{C}$. We shall obviously have to modify the definitions for, if $(\boldsymbol{p}, \boldsymbol{q})$ is in $\boldsymbol{C}$ for all $\boldsymbol{p}$, $\boldsymbol{q}$, then $(\boldsymbol{p}, \boldsymbol{p})$ is not necessarily real and the previous definition of $\|\boldsymbol{p}\|$ no longer gives a real length. We shall overcome this difficulty if we redefine an inner product on V by replacing (4), (5), and (6) by the following conditions. For all $\boldsymbol{p}, \boldsymbol{q}, \boldsymbol{r} \in V$ and all $x \in \boldsymbol{C}$, $(\boldsymbol{p}, \boldsymbol{q})$ is a mapping of pairs of elements in V into the field $\boldsymbol{C}$ satisfying

(8) $(\boldsymbol{p}, \boldsymbol{q}) = \overline{(\boldsymbol{q}, \boldsymbol{p})}$,

(9) $(\boldsymbol{p} + \boldsymbol{q}, \boldsymbol{r}) = (\boldsymbol{p}, \boldsymbol{r}) + (\boldsymbol{q}, \boldsymbol{r})$, $(x\boldsymbol{p}, \boldsymbol{q}) = x(\boldsymbol{p}, \boldsymbol{q})$,

(10) $(\boldsymbol{p}, \boldsymbol{p}) \geqslant 0$ for all $\boldsymbol{p}$,

$(\boldsymbol{p}, \boldsymbol{p}) = 0$ if and only if $\boldsymbol{p} = \boldsymbol{0}$.

$\overline{(\boldsymbol{q}, \boldsymbol{p})}$ denotes the complex conjugate of $(\boldsymbol{q}, \boldsymbol{p})$, and a mapping satisfying (8) is called *hermitian*. From (9) and (8) together we can deduce

$$\overline{(\boldsymbol{p} + \boldsymbol{q}, \boldsymbol{r})} = \overline{(\boldsymbol{p}, \boldsymbol{r})} + \overline{(\boldsymbol{q}, \boldsymbol{r})},$$

or

$$(\boldsymbol{r}, \boldsymbol{p} + \boldsymbol{q}) = (\boldsymbol{r}, \boldsymbol{p}) + (\boldsymbol{r}, \boldsymbol{q}),$$

and also

$$\begin{aligned}(\boldsymbol{p}, x\boldsymbol{q}) &= \overline{(x\boldsymbol{q}, \boldsymbol{p})} = \overline{x(\boldsymbol{q}, \boldsymbol{p})} \\ &= \bar{x} \,.\, \overline{(\boldsymbol{q}, \boldsymbol{p})} = \bar{x}(\boldsymbol{p}, \boldsymbol{q}).\end{aligned}$$

We say that $(\boldsymbol{p}, \boldsymbol{q})$ is linear in the first variable and conjugate linear in the second.

Since $(\boldsymbol{p}, \boldsymbol{p}) = \overline{(\boldsymbol{p}, \boldsymbol{p})}$ by (8), it follows that $(\boldsymbol{p}, \boldsymbol{p})$ is real for all $\boldsymbol{p}$. We postulate in (10) that $(\boldsymbol{p}, \boldsymbol{p}) \geqslant 0$ for all $\boldsymbol{p}$, and we can now define the real non-negative number $\|\boldsymbol{p}\|$, called the length of $\boldsymbol{p}$, by

$$\|\boldsymbol{p}\| = +\sqrt{(\boldsymbol{p}, \boldsymbol{p})}.$$

We then have $\|\boldsymbol{p}\| \geqslant 0$ for all $\boldsymbol{p}$ and $\|\boldsymbol{p}\| = 0$ if and only if $\boldsymbol{p} = \boldsymbol{0}$. Now

$$\|-\boldsymbol{p}\|^2 = (-\boldsymbol{p}, -\boldsymbol{p}) = (\boldsymbol{p}, \boldsymbol{p}) = \|\boldsymbol{p}\|^2,$$

so that

$$\|-\boldsymbol{p}\| = \|\boldsymbol{p}\|,$$

and

$$\|\boldsymbol{p} - \boldsymbol{q}\| = \|\boldsymbol{q} - \boldsymbol{p}\|.$$

Again, suppose $\boldsymbol{q} \neq \boldsymbol{0}$ and write $\boldsymbol{r} = \boldsymbol{q}/\|\boldsymbol{q}\|$, so that $\|\boldsymbol{r}\| = 1$. For any vector $\boldsymbol{p}$ we have

$$\begin{aligned}0 \leqslant \|\boldsymbol{p} - (\boldsymbol{p}, \boldsymbol{r})\boldsymbol{r}\|^2 &= (\boldsymbol{p} - (\boldsymbol{p}, \boldsymbol{r})\boldsymbol{r}, \boldsymbol{p} - (\boldsymbol{p}, \boldsymbol{r})\boldsymbol{r}) \\ &= (\boldsymbol{p}, \boldsymbol{p}) - (\boldsymbol{p}, \boldsymbol{r})\,(\boldsymbol{r}, \boldsymbol{p}) - \overline{(\boldsymbol{p}, \boldsymbol{r})}\,(\boldsymbol{p}, \boldsymbol{r}) + (\boldsymbol{p}, \boldsymbol{r})\,\overline{(\boldsymbol{p}, \boldsymbol{r})}\,(\boldsymbol{r}, \boldsymbol{r}) \\ &= \|\boldsymbol{p}\|^2 - 2|(\boldsymbol{p}, \boldsymbol{r})|^2 + |(\boldsymbol{p}, \boldsymbol{r})|^2\|\boldsymbol{r}\|^2 \\ &= \|\boldsymbol{p}\|^2 - |(\boldsymbol{p}, \boldsymbol{r})|^2.\end{aligned}$$

Hence

$$\begin{aligned}|(\boldsymbol{p}, \boldsymbol{r})| &\leqslant \|\boldsymbol{p}\|, \\ |(\boldsymbol{p}, \boldsymbol{q})| &\leqslant \|\boldsymbol{p}\|\;\|\boldsymbol{q}\|.\end{aligned}$$

This obviously holds for $\boldsymbol{q} = \boldsymbol{0}$ with both sides equal to zero. It is the Schwarz inequality and holds for all $\boldsymbol{p}, \boldsymbol{q} \in V$. It now follows that

$$\begin{aligned}\|\boldsymbol{p} + \boldsymbol{q}\|^2 &= (\boldsymbol{p} + \boldsymbol{q}, \boldsymbol{p} + \boldsymbol{q}) \\ &= (\boldsymbol{p}, \boldsymbol{p}) + (\boldsymbol{p}, \boldsymbol{q}) + (\boldsymbol{q}, \boldsymbol{p}) + (\boldsymbol{q}, \boldsymbol{q}) \\ &= \|\boldsymbol{p}\|^2 + (\boldsymbol{p}, \boldsymbol{q}) + \overline{(\boldsymbol{p}, \boldsymbol{q})} + \|\boldsymbol{q}\|^2 \\ &= \|\boldsymbol{p}\|^2 + 2\mathrm{Re}(\boldsymbol{p}, \boldsymbol{q}) + \|\boldsymbol{q}\|^2 \\ &\leqslant \|\boldsymbol{p}\|^2 + 2|(\boldsymbol{p}, \boldsymbol{q})| + \|\boldsymbol{q}\|^2 \\ &\leqslant \|\boldsymbol{p}\|^2 + 2\|\boldsymbol{p}\|\;\|\boldsymbol{q}\| + \|\boldsymbol{q}\|^2 \\ &= \{\|\boldsymbol{p}\| + \|\boldsymbol{q}\|\}^2,\end{aligned}$$

and hence $\|\boldsymbol{p} + \boldsymbol{q}\| \leqslant \|\boldsymbol{p}\| + \|\boldsymbol{q}\|$.

This is the triangle inequality, and we have shown that $d(\boldsymbol{p}, \boldsymbol{q}) = \|\boldsymbol{p} - \boldsymbol{q}\|$ satisfies the conditions for a distance function, or metric.

Definition. A vector space V over the complex field $\boldsymbol{C}$, on which an inner product is defined, is called a *unitary space.*

By analogy with euclidean space, we say that two vectors $\boldsymbol{p}, \boldsymbol{q}$ in the unitary space V are orthogonal if and only if $(\boldsymbol{p}, \boldsymbol{q}) = 0$. In this case $(\boldsymbol{q}, \boldsymbol{p}) = 0$ also. Suppose that we can choose an orthonormal basis for V, i.e., a basis $\{\boldsymbol{e}_1, \boldsymbol{e}_2, \ldots, \boldsymbol{e}_n\}$ such that $\|\boldsymbol{e}_k\| = 1$ $(k = 1, 2, \ldots, n)$ and $(\boldsymbol{e}_h, \boldsymbol{e}_k) = 0$ $(h \neq k)$. These conditions can be expressed by the single equation

$$(\boldsymbol{e}_h, \boldsymbol{e}_k) = \delta_{hk} \qquad (h = 1, \ldots, n;\ k = 1, \ldots, n),$$

where $\delta_{h,k}$ is the Kronecker delta. Let the vector $\boldsymbol{p}$ have coordinate vector $(x_1, x_2, \ldots, x_n)$ relative to this basis. Then

$$\begin{aligned}
\|\boldsymbol{p}\|^2 &= (x_1\boldsymbol{e}_1 + \cdots + x_n\boldsymbol{e}_n, x_1\boldsymbol{e}_1 + \cdots + x_n\boldsymbol{e}_n) \\
&= \sum_{h,k} x_h\bar{x}_k(\boldsymbol{e}_h, \boldsymbol{e}_k) \\
&= \sum_{k=1}^{n} x_k\bar{x}_k \\
&= |x_1|^2 + |x_2|^2 + \cdots + |x_n|^2.
\end{aligned}$$

If $\boldsymbol{q}$ has coordinate vector $(y_1, y_2, \ldots, y_n)$, then

$$\begin{aligned}
(\boldsymbol{p}, \boldsymbol{q}) &= \sum_{h,k} x_h\bar{y}_k(\boldsymbol{e}_h, \boldsymbol{e}_k) \\
&= x_1\bar{y}_1 + x_2\bar{y}_2 + \cdots + x_n\bar{y}_n.
\end{aligned}$$

Note that conditions (4), (5), and (6) for an inner product on a euclidean space are contained in conditions (8), (9), and (10) for an inner product on a unitary space. For if $(\boldsymbol{p}, \boldsymbol{q})$ is real for each $\boldsymbol{p}, \boldsymbol{q}$ then $\overline{(\boldsymbol{q}, \boldsymbol{p})} = (\boldsymbol{q}, \boldsymbol{p})$.

We now show that every unitary space of finite dimension has an orthonormal basis.

7.6 Orthonormal bases

Let V be a unitary space and let $\{\boldsymbol{v}_1, \boldsymbol{v}_2, \ldots, \boldsymbol{v}_n\}$ be a basis for V. We first construct an orthogonal set $\{\boldsymbol{w}_1, \boldsymbol{w}_2, \ldots, \boldsymbol{w}_n\}$ of non-zero vectors with the properties

(1) $(\boldsymbol{w}_h, \boldsymbol{w}_k) = 0$ $(h \neq k)$,

(2) $\boldsymbol{w}_k$ is a linear combination of $\boldsymbol{v}_1, \ldots, \boldsymbol{v}_k$ $\qquad (k = 1, \ldots, n)$.

The basis vector $\boldsymbol{v}_1$ is non-zero and we take $\boldsymbol{w}_1 = \boldsymbol{v}_1$. Write $\boldsymbol{w}_2 = \boldsymbol{v}_2 - \lambda \boldsymbol{v}_1$, where $\lambda \in \boldsymbol{C}$. Now

$$
\begin{aligned}
(\boldsymbol{w}_2, \boldsymbol{w}_1) &= (\boldsymbol{v}_2 - \lambda \boldsymbol{v}_1, \boldsymbol{v}_1) \\
&= (\boldsymbol{v}_2, \boldsymbol{v}_1) - \lambda(\boldsymbol{v}_1, \boldsymbol{v}_1),
\end{aligned}
$$

so that $(\boldsymbol{w}_2, \boldsymbol{w}_1) = 0$ if and only if $\lambda = (\boldsymbol{v}_2, \boldsymbol{v}_1)/\|\boldsymbol{v}_1\|^2$. $\|\boldsymbol{v}_1\|^2 \neq 0$ since $\boldsymbol{v}_1 \neq \boldsymbol{0}$, and $\boldsymbol{w}_2 \neq \boldsymbol{0}$ since it is a non-trivial linear combination of the linearly independent vectors $\boldsymbol{v}_1, \boldsymbol{v}_2$. We now proceed by induction.

Suppose that $\{\boldsymbol{w}_1, \ldots, \boldsymbol{w}_k\}$ have been constructed with the properties (1) and (2). Let

$$
\boldsymbol{w}_{k+1} = \boldsymbol{v}_{k+1} - \sum_{h=1}^{k} \lambda_h \boldsymbol{w}_h.
$$

Now if $1 \leqslant j \leqslant k$

$$
\begin{aligned}
(\boldsymbol{w}_{k+1}, \boldsymbol{w}_j) &= (\boldsymbol{v}_{k+1}, \boldsymbol{w}_j) - \sum_{h=1}^{k} \lambda_h (\boldsymbol{w}_h, \boldsymbol{w}_j) \\
&= (\boldsymbol{v}_{k+1}, \boldsymbol{w}_j) - \lambda_j (\boldsymbol{w}_j, \boldsymbol{w}_j).
\end{aligned}
$$

$\boldsymbol{w}_j \neq \boldsymbol{0}$ and so $\|\boldsymbol{w}_j\| \neq 0$. Hence $(\boldsymbol{w}_{k+1}, \boldsymbol{w}_j) = 0$ $(j = 1, \ldots, k)$ if and only if

$$
\lambda_j = (\boldsymbol{v}_{k+1}, \boldsymbol{w}_j)/\|\boldsymbol{w}_j\|^2 \qquad (j = 1, \ldots, k).
$$

$\boldsymbol{w}_{k+1}$ is now uniquely determined. $\boldsymbol{w}_h$ is a linear combination of $\boldsymbol{v}_1, \ldots, \boldsymbol{v}_h$ $(h = 1, \ldots, k)$ so that $\sum_{h=1}^{k} \lambda_h \boldsymbol{w}_h$ is a linear combination of $\boldsymbol{v}_1, \ldots, \boldsymbol{v}_k$, and $\boldsymbol{w}_{k+1}$ is a non-trivial linear combination of $\boldsymbol{v}_1, \ldots, \boldsymbol{v}_{k+1}$. $\boldsymbol{w}_{k+1} \neq \boldsymbol{0}$, since $\boldsymbol{v}_1, \ldots, \boldsymbol{v}_{k+1}$ are linearly independent. We have now constructed a set $\{\boldsymbol{w}_1, \ldots, \boldsymbol{w}_{k+1}\}$ having properties (1) and (2). But we showed how to construct $\{\boldsymbol{w}_1, \boldsymbol{w}_2\}$ having these properties and it follows by induction that we can construct an orthogonal set $\{\boldsymbol{w}_1, \boldsymbol{w}_2, \ldots, \boldsymbol{w}_n\}$ as required.

We now show that this set is linearly independent and hence forms a basis for V. Suppose that

$$
\alpha_1 \boldsymbol{w}_1 + \alpha_2 \boldsymbol{w}_2 + \cdots + \alpha_n \boldsymbol{w}_n = \boldsymbol{0} \qquad (\alpha_1, \ldots, \alpha_n \in \boldsymbol{C}).
$$

Taking the inner product of the left-hand side with $\boldsymbol{w}_k$ gives

$$
\alpha_k (\boldsymbol{w}_k, \boldsymbol{w}_k) = 0 \qquad (k = 1, \ldots, n),
$$

i.e.,

$$
\alpha_k \|\boldsymbol{w}_k\|^2 = 0.
$$

But $\|\boldsymbol{w}_k\| \neq 0$, and hence $\alpha_k = 0$ $(k = 1, \ldots, n)$. Thus $\boldsymbol{w}_1, \boldsymbol{w}_2, \ldots, \boldsymbol{w}_n$ are linearly independent. Now take $\boldsymbol{e}_k = \boldsymbol{w}_k/\|\boldsymbol{w}_k\|$ $(k = 1, \ldots, n)$, so that $\boldsymbol{e}_1, \boldsymbol{e}_2, \ldots, \boldsymbol{e}_n$ are also linearly independent. Then

$$(\boldsymbol{e}_h, \boldsymbol{e}_k) = \frac{(\boldsymbol{w}_h, \boldsymbol{w}_k)}{\|\boldsymbol{w}_h\| \, \|\boldsymbol{w}_k\|} = \delta_{h,k}.$$

$\{\boldsymbol{e}_1, \boldsymbol{e}_2, \ldots, \boldsymbol{e}_n\}$ is therefore an orthonormal basis for V.

We have shown that, given any basis of V, we can construct an orthonormal basis for V. But every vector space of finite dimension has a basis. We have therefore proved that every unitary space of finite dimension has an orthonormal basis. The above process is known as the *Schmidt orthogonalization process.*

The proof holds exactly as its stands for a euclidean space, so that every euclidean space of finite dimension has an orthonormal basis.

Example 1. M is the subspace of $\boldsymbol{R}_4$ generated by $\{1, 0, 1, 0\}$, $\{1, 2, 0, -3\}$, and $\{0, -1, 5, 2\}$ and the inner product of two vectors $\boldsymbol{x} = \{x_1, x_2, x_3, x_4\}$ and $\boldsymbol{y} = \{y_1, y_2, y_3, y_4\}$ is defined to be

$$(\boldsymbol{x}, \boldsymbol{y}) = x_1y_1 + x_2y_2 + x_3y_3 + x_4y_4.$$

Find an orthonormal basis for M.

The definition of $(\boldsymbol{x}, \boldsymbol{y})$ clearly satisfies the conditions for an inner product. Take $\boldsymbol{w}_1 = \{1, 0, 1, 0\}$ and let

$$\begin{aligned} \boldsymbol{u}_2 &= \{1, 2, 0, -3\} - \lambda\{1, 0, 1, 0\} \\ &= \{1 - \lambda, 2, -\lambda, -3\}. \\ (\boldsymbol{u}_2, \boldsymbol{w}_1) &= 1 - \lambda - \lambda = 0 \text{ if } \lambda = \tfrac{1}{2}. \end{aligned}$$

Thus $$\boldsymbol{u}_2 = \{\tfrac{1}{2}, 2, -\tfrac{1}{2}, -3\}.$$

We can take $\boldsymbol{w}_2 = k\boldsymbol{u}_2$, where k is any scalar, since $(k\boldsymbol{u}_2, \boldsymbol{w}_1) = k(\boldsymbol{u}_2, \boldsymbol{w}_1) = 0$, and it is convenient here to take $k = 2$, so that

$$\boldsymbol{w}_2 = \{1, 4, -1, -6\}.$$

Now let $$\boldsymbol{u}_3 = \{0, -1, 5, 2\} - \mu\boldsymbol{w}_1 - \nu\boldsymbol{w}_2.$$

Then $$\begin{aligned} (\boldsymbol{u}_3, \boldsymbol{w}_1) &= 5 - \mu(1 + 1) = 0, \\ (\boldsymbol{u}_3, \boldsymbol{w}_2) &= -4 - 5 - 12 - \nu(1 + 16 + 1 + 36) = 0. \end{aligned}$$

Hence $$\mu = 5/2, \qquad \nu = -21/54 = -7/18.$$

Therefore $\boldsymbol{u}_3 = \{0, -1, 5, 2\} - \tfrac{5}{2}\{1, 0, 1, 0\} + \tfrac{7}{18}\{1, 4, -1, 6\}$

and we may take

$$\boldsymbol{w}_3 = 9\boldsymbol{u}_3 = \{0, -9, 45, 18\} - \tfrac{1}{2}\{45, 0, 45, 0\} + \tfrac{1}{2}\{7, 28, -7, -42\}$$
$$= \{-19, 5, 19, -3\}.$$

$\boldsymbol{w}_1$, $\boldsymbol{w}_2$, $\boldsymbol{w}_3$ form an orthogonal basis for M, and we now normalize these vectors to give an orthonormal basis

$$\boldsymbol{e}_1 = \frac{1}{\sqrt{2}}\{1, 0, 1, 0\}, \qquad \boldsymbol{e}_2 = \frac{1}{\sqrt{54}}\{1, 4, -1, -6\},$$

$$\boldsymbol{e}_3 = \frac{1}{\sqrt{756}}\{-19, 5, 19, -3\} = \frac{1}{6\sqrt{21}}\{-19, 5, 19, -3\}.$$

7.7 Matrix of an inner product

Let $\{\boldsymbol{v}_1, \boldsymbol{v}_2, \ldots, \boldsymbol{v}_n\}$ be any ordered basis of a vector space V with an inner product. Let

$$(\boldsymbol{v}_i, \boldsymbol{v}_j) = a_{ij}.$$

The $n \times n$ matrix $\boldsymbol{A}$ with elements a_{ij} is called the *matrix of the inner product with respect to the given ordered basis*. Let $\boldsymbol{p}$, $\boldsymbol{q}$ have coordinate vectors $\{x_1, x_2, \ldots, x_n\}$, $\{y_1, y_2, \ldots, y_n\}$ relative to this basis. Then, if V is a euclidean space,

$$\begin{aligned}(\boldsymbol{p}, \boldsymbol{q}) &= (x_1\boldsymbol{v}_1 + \cdots + x_n\boldsymbol{v}_n, y_1\boldsymbol{v}_1 + \cdots + y_n\boldsymbol{v}_n) \\ &= \sum_{h,k} x_h y_k(\boldsymbol{v}_h, \boldsymbol{v}_k) \\ &= \sum_{h,k} x_h a_{hk} y_k.\end{aligned}$$

In matrix notation this can be written

$$(\boldsymbol{p}, \boldsymbol{q}) = \boldsymbol{x}^{\mathrm{t}}\boldsymbol{A}\boldsymbol{y},$$

where $\boldsymbol{x}$, $\boldsymbol{y}$ are the (column) coordinate vectors of $\boldsymbol{p}$, $\boldsymbol{q}$ respectively. The matrix $\boldsymbol{A}$ depends on the particular basis chosen. If the basis is orthonormal, then $\boldsymbol{A}$ is the unit matrix, but if the basis is not orthogonal, then $\boldsymbol{A}$ is not even diagonal. We therefore investigate the effect on the matrix $\boldsymbol{A}$ of changing the basis of V. Let us change to a new basis with respect to which $\boldsymbol{p}$ and $\boldsymbol{q}$ have coordinate vectors ξ and η respectively. Then $\boldsymbol{x} = \boldsymbol{P}\xi$, $\boldsymbol{y} = \boldsymbol{P}\eta$, where $\boldsymbol{P}$ is a non-singular $n \times n$ matrix (see Section 3.6). Then

$$\begin{aligned}(\boldsymbol{p}, \boldsymbol{q}) &= (\boldsymbol{P}\xi)^{\mathrm{t}}\boldsymbol{A}(\boldsymbol{P}\eta) \\ &= \xi^{\mathrm{t}}\boldsymbol{P}^{\mathrm{t}}\boldsymbol{A}\boldsymbol{P}\eta = \xi^{\mathrm{t}}\boldsymbol{B}\eta,\end{aligned}$$

where $\boldsymbol{B} = \boldsymbol{P}^t\boldsymbol{A}\boldsymbol{P}$. Hence the matrix of the inner product with respect to the new basis is $\boldsymbol{P}^t\boldsymbol{A}\boldsymbol{P}$. Two square matrices $\boldsymbol{A}$ and $\boldsymbol{B}$ of the same order are said to be *congruent* if $\boldsymbol{B} = \boldsymbol{P}^t\boldsymbol{A}\boldsymbol{P}$ for some non-singular matrix $\boldsymbol{P}$. Note that in this case $\boldsymbol{A} = (\boldsymbol{P}^{-1})^t\boldsymbol{B}\boldsymbol{P}^{-1} = \boldsymbol{Q}^t\boldsymbol{B}\boldsymbol{Q}$ where $\boldsymbol{Q} = \boldsymbol{P}^{-1}$ is non-singular. Thus the matrices of a symmetric inner product with respect to two different bases of V are congruent.

When the space is euclidean, $(\boldsymbol{p}, \boldsymbol{q}) = (\boldsymbol{q}, \boldsymbol{p})$ for every $\boldsymbol{p}, \boldsymbol{q} \in V$. In particular

$$(\boldsymbol{v}_i, \boldsymbol{v}_j) = (\boldsymbol{v}_j, \boldsymbol{v}_i) \text{ for each } i, j;$$

i.e.,

$$a_{ij} = a_{ji},$$

or

$$\boldsymbol{A} = \boldsymbol{A}^t.$$

A matrix $\boldsymbol{A}$ for which $\boldsymbol{A}^t = \boldsymbol{A}$ is called *symmetric* and the matrix of a symmetric inner product is therefore symmetric.

When the space is unitary, $(\boldsymbol{p}, \boldsymbol{q}) = \overline{(\boldsymbol{q}, \boldsymbol{p})}$ for all $\boldsymbol{p}, \boldsymbol{q} \in V$. In this case

$$(\boldsymbol{p}, \boldsymbol{q}) = \boldsymbol{x}^t\boldsymbol{A}\overline{\boldsymbol{y}} = \xi^t\boldsymbol{P}^t\boldsymbol{A}\overline{\boldsymbol{P}\eta}, \text{ so that } \boldsymbol{B} = \boldsymbol{P}^t\boldsymbol{A}\overline{\boldsymbol{P}}.$$

$$(\boldsymbol{v}_i, \boldsymbol{v}_j) = \overline{(\boldsymbol{v}_j, \boldsymbol{v}_i)} \text{ for each } i, j;$$

i.e.,

$$a_{ij} = \overline{a}_{ji},$$

or

$$\boldsymbol{A} = \overline{\boldsymbol{A}}^t.$$

A matrix $\boldsymbol{A}$ for which $\overline{\boldsymbol{A}}^t = \boldsymbol{A}$ is called *hermitian* and the matrix of a hermitian inner product is therefore hermitian.

7.8 Orthogonal complements

Let M be a subspace of an inner product space V and let $M^\perp$ (read M perp) be the set of all vectors $\boldsymbol{q}$ in V such that $(\boldsymbol{p}, \boldsymbol{q}) = 0$ for all $\boldsymbol{p} \in M$. This implies that $(\boldsymbol{q}, \boldsymbol{p}) = 0$ also. Clearly M is a subspace of V. If any two subspaces M, N of V are such that every vector of M is orthogonal to every vector of N, we say that M and N are *orthogonal subspaces.*

THEOREM 7.1. If M is a subspace of an inner product space V, then $V = M \oplus M^\perp$ and $M^{\perp\perp} = M$.

Proof. Let $\dim V = n$, $\dim M = m$. If $m = n$, then $M = V$ and $M^\perp$ is the set of all $\boldsymbol{q}$ such that $(\boldsymbol{p}, \boldsymbol{q}) = 0$ for all $\boldsymbol{p} \in V$. Let $\boldsymbol{p}, \boldsymbol{q}$ have coordinate vectors $\boldsymbol{x}, \boldsymbol{y}$ relative to an orthonormal basis of V. Then $\boldsymbol{x}^t\boldsymbol{y} = 0$ for all $\boldsymbol{x}$.

This implies that $\boldsymbol{y} = \boldsymbol{0}$, and hence $\boldsymbol{q} = \boldsymbol{0}$. Thus M consists of the zero vector alone and $M \oplus M^{\perp} = M = V$. Again $M^{\perp\perp}$, the set of all vectors in V that are orthogonal to the zero vector, is V, so that $M^{\perp\perp} = V = M$. The theorem is therefore true if $m = n$.

Now suppose $m < n$. Let $\{\boldsymbol{e}_1, \boldsymbol{e}_2, \ldots, \boldsymbol{e}_m\}$ be an orthonormal basis of M and extend this to an orthonormal basis $\{\boldsymbol{e}_1, \boldsymbol{e}_2, \ldots, \boldsymbol{e}_m, \boldsymbol{e}_{m+1}, \ldots, \boldsymbol{e}_n\}$ of V. Let $\boldsymbol{q} \in M^{\perp}$ have coordinate vector $\{y_1, y_2, \ldots, y_n\}$ relative to this basis. Then

$$(\boldsymbol{q}, \boldsymbol{e}_k) = 0 \qquad (k = 1, \ldots, m),$$

so that

$$\sum_{i=1}^{n} y_i(\boldsymbol{e}_i, \boldsymbol{e}_k) = 0 \qquad (k = 1, \ldots, m),$$

or

$$y_k = 0 \qquad (k = 1, \ldots, m).$$

It follows that $\boldsymbol{q}$ is contained in the set spanned by $\boldsymbol{e}_{m+1}, \ldots, \boldsymbol{e}_n$. But each of these vectors is orthogonal to each of the vectors $\boldsymbol{e}_1, \boldsymbol{e}_2, \ldots, \boldsymbol{e}_m$ and hence to M, so that $M^{\perp}$ is the set $[\boldsymbol{e}_{m+1}, \ldots, \boldsymbol{e}_n]$. These vectors of the spanning set are orthonormal and so are linearly independent, and $\dim M = n - m$. Thus

$$\dim M + \dim M^{\perp} = m + (n - m) = n = \dim V.$$

But if $\boldsymbol{p} \in M \cap M^{\perp}$, then $(\boldsymbol{p}, \boldsymbol{p}) = 0$, so that $\boldsymbol{p} = \boldsymbol{0}$ and $M \cap M^{\perp} = \boldsymbol{0}$. Hence $V = M \oplus M^{\perp}$.

Now $M^{\perp\perp}$, the set of all vectors orthogonal to $M^{\perp}$, certainly includes M. But, by the above result applied to $M^{\perp}$ in place of M,

$$\begin{aligned} \dim M^{\perp\perp} &= n - \dim M^{\perp} \\ &= n - (n - \dim M) \\ &= \dim M. \end{aligned}$$

Hence $M^{\perp\perp} = M$ and the theorem is now proved.

Definition. $M^{\perp}$ is called the *orthogonal complement* of M. Obviously M is also the orthogonal complement of $M^{\perp}$. We note that, since $V = M \oplus M^{\perp}$, any vector $\boldsymbol{v} \in V$ can be expressed uniquely in the form $\boldsymbol{p} + \boldsymbol{q}$, where $\boldsymbol{p} \in M$, $\boldsymbol{q} \in M^{\perp}$. The vector $\boldsymbol{p}$ is called the projection of $\boldsymbol{v}$ on M. In particular, in a euclidean space of dimension 3, any vector can be expressed uniquely as the sum of a vector in a given plane (its projection on the plane) and a vector perpendicular to the plane.

Example 2. M is the subspace of $\boldsymbol{R}_4$ generated by the vectors $\{0, 1, 0, 1\}$ and $\{2, 0, -3, -1\}$. With the standard inner product, find an orthogonal basis for $M^\perp$. Find the projection on M of the vector $\{1, 1, 1, 1\}$.

Write $\boldsymbol{u}_1 = \{0, 1, 0, 1\}$ and $\boldsymbol{u}_2 = \{2, 0, -3, -1\}$. Every vector $\boldsymbol{x} = \{x_1, x_2, x_3, x_4\}$ in $M^\perp$ satisfies the conditions $(\boldsymbol{x}, \boldsymbol{u}_1) = 0$, $(\boldsymbol{x}, \boldsymbol{u}_2) = 0$. Thus $x_2 + x_4 = 0$ and $2x_1 - 3x_3 - x_4 = 0$.

Writing $x_1 = 3h$, $x_2 = 3k$, we see that every vector in $M^\perp$ is of the form $\{3h, 3k, 2h + k, -3k\}$. This can be written $h\{3, 0, 2, 0\} + k\{0, 3, 1, -3\}$, so that $M^\perp$ has a basis $\{3, 0, 2, 0\}$, $\{0, 3, 1, -3\}$. Write

$$\begin{aligned} \boldsymbol{v}_1 &= \{3, 0, 2, 0\}, \\ \boldsymbol{w} &= \{3, 0, 2, 0\} - \lambda\{0, 3, 1, -3\} \\ &= \{3, -3\lambda, 2 - \lambda, 3\lambda\}. \\ (\boldsymbol{v}_1, \boldsymbol{w}) &= 9 + 4 - 2\lambda = 0 \quad \text{if} \quad \lambda = 13/2. \end{aligned}$$

Take $\boldsymbol{v}_2 = 2\boldsymbol{w} = \{6, -39, -9, 39\}$. Then $(\boldsymbol{v}_1, \boldsymbol{v}_2) = 0$ and $\{\boldsymbol{v}_1, \boldsymbol{v}_2\}$ is an orthogonal basis for $M^\perp$. Clearly $\{\boldsymbol{u}_1, \boldsymbol{u}_2, \boldsymbol{v}_1, \boldsymbol{v}_2\}$ is a basis for $\boldsymbol{R}_4$ and we may express $\{1, 1, 1, 1\}$ in the form $\alpha\boldsymbol{u}_1 + \beta\boldsymbol{u}_2 + \gamma\boldsymbol{v}_1 + \delta\boldsymbol{v}_2$. Taking the inner product with $\boldsymbol{u}_1$, $\boldsymbol{u}_2$ in turn we obtain

$$\begin{aligned} 2 &= \alpha(\boldsymbol{u}_1, \boldsymbol{u}_1) + \beta(\boldsymbol{u}_2, \boldsymbol{u}_1) = 2\alpha + (-1)\beta, \\ -2 &= \alpha(\boldsymbol{u}_1, \boldsymbol{u}_2) + \beta(\boldsymbol{u}_2, \boldsymbol{u}_2) = -\alpha + 14\beta. \end{aligned}$$

Thus $\alpha = 26/27$, $\beta = -2/27$ and the projection of $\{1, 1, 1, 1\}$ on M is

$$\alpha\boldsymbol{u}_1 + \beta\boldsymbol{u}_2 = \tfrac{1}{27}\{-4, 26, 6, 28\}.$$

Example 3. Relative to the standard basis the matrix of a hermitian inner product on the space $\boldsymbol{C}_3$ is

$$\boldsymbol{A} = \begin{bmatrix} 2 & i & 0 \\ -i & 1 & 0 \\ 0 & 0 & 3 \end{bmatrix}.$$

The subspace M is generated by the vector $\{1, i, -1\}$. Find an orthogonal basis for $M^\perp$.

The vector $\boldsymbol{x} = \{x_1, x_2, x_3\}$ is in $M^\perp$ if

$$[x_1\ x_2\ x_3]\begin{bmatrix} 2 & i & 0 \\ -i & 1 & 0 \\ 0 & 0 & 3 \end{bmatrix}\begin{bmatrix} 1 \\ -i \\ -1 \end{bmatrix} = 0,$$

i.e.,

$$3x_1 - 2ix_2 - 3x_3 = 0.$$

Writing $x_1 = h$, $x_2 = 3k$, we see that $M^\perp$ consists of all vectors of the form $\{h, 3k, h - 2ik\}$. The vectors $\boldsymbol{u} = \{1, 0, 1\}$ and $\boldsymbol{v} = \{0, 3, -2i\}$ form a basis for $M^\perp$.

$$\boldsymbol{u}^t A\overline{\boldsymbol{u}} = 5 \qquad \text{and} \qquad \boldsymbol{v}^t A\overline{\boldsymbol{u}} = -9i.$$

Thus
$$(5\boldsymbol{v} + 9i\boldsymbol{u})^t A\overline{\boldsymbol{u}} = 0$$

and the vector $\boldsymbol{w} = 5\boldsymbol{v} + 9i\boldsymbol{u} = (9i, 15, -i)$ is orthogonal to $\boldsymbol{u}$. The vectors $\boldsymbol{u}$ and $\boldsymbol{w}$ form an orthogonal basis for $M^\perp$.

Problems

7.1 In a euclidean or unitary space prove that

$$\|\boldsymbol{u} - \boldsymbol{v}\|^2 + \|\boldsymbol{u} + \boldsymbol{v}\|^2 = 2\|\boldsymbol{u}\|^2 + 2\|\boldsymbol{v}\|^2.$$

Interpret this result geometrically in the space $\boldsymbol{R}_2$ and hence discover why this equation is known as the parallelogram equation.

7.2 Prove that

$$(\boldsymbol{x}, \boldsymbol{y}) = x_1y_1 + x_1y_2 + x_2y_1 + 2x_2y_2$$

defines an inner product in the space $\boldsymbol{R}_2$.

7.3 In an inner product space prove that, if $\|\boldsymbol{u}\| = \|\boldsymbol{v}\|$, then $\boldsymbol{u} - \boldsymbol{v}$ is orthogonal to $\boldsymbol{u} + \boldsymbol{v}$. Interpret this result geometrically in the space $\boldsymbol{R}_2$.

7.4 In the space $\boldsymbol{R}_3$, with the standard inner product, find an orthogonal basis for the orthogonal complement of the subspace M generated by $\{2, 1, -1\}$. Express the vector $\{1, 0, 3\}$ in the form $\boldsymbol{p} + \boldsymbol{q}$, where $\boldsymbol{p} \in M$, $\boldsymbol{q} \in M^\perp$.

7.5 V is the vector space over $\boldsymbol{R}$ whose elements are the continuous functions $x(t)$ on the interval $0 \leqslant t \leqslant 1$. Prove that we can convert V into a euclidean space by defining an inner product

$$(\boldsymbol{x}, \boldsymbol{y}) = \int_0^1 x(t)y(t)dt.$$

Prove that $\cos 2m\pi t$ and $\cos 2n\pi t$ $(m \neq n)$ are orthogonal, and find a quadratic polynomial that is orthogonal to both 1 and t.

7.6 If M and N are subspaces of a finite-dimensional inner product space, prove that

$$(M + N)^\perp = M^\perp \cap N^\perp$$

and

$$(M \cap N)^\perp = M^\perp + N^\perp.$$

7.7 $\boldsymbol{u}$ and $\boldsymbol{v}$ are two fixed non-zero vectors of a euclidean vector space. Find the shortest vector of the form $\boldsymbol{w} = \boldsymbol{u} + \lambda\boldsymbol{v}$ and show that it is orthogonal to $\boldsymbol{v}$. Interpret this geometrically in the space $\boldsymbol{R}_2$.

7.8 Find an orthogonal basis for the subspace M of the euclidean space $\boldsymbol{R}_4$ (with the standard inner product) spanned by the vectors

$\{1, 2, 2, 0\}$, $\{0, 1, 5, 4\}$, and $\{1, 1, 3, 4\}$. Find $M^{\perp}$. Find the projection of the vector $\{5, -11, 13, 9\}$ on M.

7.9 Relative to the standard basis the matrix of a hermitian inner product on the space C_3 is

$$A = \begin{bmatrix} 1 & 0 & 1+i \\ 0 & 2 & 0 \\ 1-i & 0 & 3 \end{bmatrix}.$$

The subspace M is generated by the vectors $\{1, 0, 0\}$ and $\{0, 1, i\}$. Find $M^{\perp}$ and find the projection on M of the vector $\{1, i, 1\}$.

7.10 Prove that $(A, B) = \operatorname{tr}(AB^{t})$ defines an inner product on the vector space of all real 2×2 matrices over the real field (see Section 2.2).

7.11 With the notation of Section 7.1, equation (7), prove that $|(p, q)| = \|p\|\,\|q\|$ if and only if p and q are linearly dependent.

If u, v, w are linearly independent vectors in the euclidean space V, prove that it is always possible to find a vector of the form $w + \lambda u + \mu v$ that is orthogonal to both u and v.

8 · *Orthogonal and Unitary Transformations*

In this chapter we consider transformations on an inner product space that have the important property of leaving distance invariant. Such transformations are called *isometries*. In three-dimensional geometry rotations and reflections are isometric linear transformations and we now generalize these concepts.

8.1 Isometries

A linear transformation T on an inner product space V is said to be an *isometry* if

$$(T(\boldsymbol{p}), T(\boldsymbol{q})) = (\boldsymbol{p}, \boldsymbol{q}) \text{ for all } \boldsymbol{p}, \boldsymbol{q} \in V.$$

This implies that $\quad \|T(\boldsymbol{p}) - T(\boldsymbol{q})\| = \|\boldsymbol{p} - \boldsymbol{q}\|.$

An isometry on a euclidean space is called an *orthogonal transformation*. Let $\boldsymbol{x}, \boldsymbol{y}$ be the coordinate vectors of $\boldsymbol{p}, \boldsymbol{q}$ relative to an orthonormal basis of V, so that

$$(\boldsymbol{p}, \boldsymbol{q}) = \boldsymbol{x}^{\mathrm{t}}\boldsymbol{y}.$$

Let $\boldsymbol{A}$ be the matrix of the orthogonal transformation T relative to the same orthonormal basis. Then

$$(T(\boldsymbol{p}), T(\boldsymbol{q})) = (\boldsymbol{A}\boldsymbol{x})^{\mathrm{t}}(\boldsymbol{A}\boldsymbol{y}) = \boldsymbol{x}^{\mathrm{t}}\boldsymbol{A}^{\mathrm{t}}\boldsymbol{A}\boldsymbol{y},$$

so that $\quad \boldsymbol{x}^{\mathrm{t}}\boldsymbol{A}^{\mathrm{t}}\boldsymbol{A}\boldsymbol{y} = \boldsymbol{x}^{\mathrm{t}}\boldsymbol{y}$ for all $\boldsymbol{x}, \boldsymbol{y}$.

Thus $\quad \boldsymbol{x}^{\mathrm{t}}(\boldsymbol{A}^{\mathrm{t}}\boldsymbol{A} - \boldsymbol{I})\boldsymbol{y} = 0$ for all $\boldsymbol{x}, \boldsymbol{y}$.

It follows that $\boldsymbol{A}^{\mathrm{t}}\boldsymbol{A} = \boldsymbol{I}$, so that $\boldsymbol{A}^{\mathrm{t}} = \boldsymbol{A}^{-1}$ and $\boldsymbol{A}\boldsymbol{A}^{\mathrm{t}} = \boldsymbol{I}$ also. A square matrix $\boldsymbol{A}$ with real elements having the property that $\boldsymbol{A}^{\mathrm{t}}\boldsymbol{A} = \boldsymbol{I}$ is called an *orthogonal matrix*. We have just shown that, relative to an orthonormal basis, the matrix of an orthogonal mapping is orthogonal.

An isometry on a unitary space is called a *unitary transformation*. Define $\boldsymbol{p}$, $\boldsymbol{q}$, $\boldsymbol{x}$, $\boldsymbol{y}$ as above and let $\boldsymbol{A}$ be the matrix of a unitary transformation relative to the same orthonormal basis. Then

$$(T(\boldsymbol{p}), T(\boldsymbol{q})) = (\boldsymbol{A}\boldsymbol{x})^{\mathrm{t}}\overline{(\boldsymbol{A}\boldsymbol{y})} = \boldsymbol{x}^{\mathrm{t}}\boldsymbol{A}^{\mathrm{t}}\overline{\boldsymbol{A}\boldsymbol{y}},$$

$$(\boldsymbol{p}, \boldsymbol{q}) = \boldsymbol{x}^{\mathrm{t}}\overline{\boldsymbol{y}},$$

and

$$\boldsymbol{x}^{\mathrm{t}}(\boldsymbol{A}^{\mathrm{t}}\overline{\boldsymbol{A}} - \boldsymbol{I})\overline{\boldsymbol{y}} = 0 \text{ for all } \boldsymbol{x}, \boldsymbol{y}.$$

Thus $\boldsymbol{A}^{\mathrm{t}}\overline{\boldsymbol{A}} = \boldsymbol{I}$ or, taking the transpose of this equation,

$$\overline{\boldsymbol{A}}^{\mathrm{t}}\boldsymbol{A} = \boldsymbol{I} = \boldsymbol{A}\overline{\boldsymbol{A}}^{\mathrm{t}}.$$

A square matrix $\boldsymbol{A}$ having the property that $\overline{\boldsymbol{A}}^{\mathrm{t}}\boldsymbol{A} = \boldsymbol{I}$ is called a *unitary matrix*.

Relative to an orthonormal basis, the matrix of a unitary mapping is unitary. A unitary matrix with real elements is an orthogonal matrix.

THEOREM 8.1. A matrix $\boldsymbol{A}$ is the matrix of a unitary (orthogonal) mapping relative to an orthonormal basis if and only if its rows form an orthonormal set of coordinate vectors.

The proof is easy and is left to the reader.

Example 1. Show that, for any real value of θ, the matrix

$$\boldsymbol{A} = \begin{bmatrix} \cos\theta & -\sin\theta \\ \sin\theta & \cos\theta \end{bmatrix}$$

is orthogonal.

$$\boldsymbol{A}^{\mathrm{t}}\boldsymbol{A} = \begin{bmatrix} \cos\theta & \sin\theta \\ -\sin\theta & \cos\theta \end{bmatrix} \begin{bmatrix} \cos\theta & -\sin\theta \\ \sin\theta & \cos\theta \end{bmatrix} = \begin{bmatrix} 1 & 0 \\ 0 & 1 \end{bmatrix} = \boldsymbol{I}.$$

The corresponding linear transformation is a rotation in a plane through an angle θ (see, e.g., Ref. 12, p. 35).

THEOREM 8.2. The eigenvalues of a unitary mapping have modulus 1.

Proof. Let λ be an eigenvalue of the unitary mapping T and let $\boldsymbol{u}$ be a corresponding eigenvector. Then

$$T\boldsymbol{u} = \lambda\boldsymbol{u}, \qquad \text{where} \qquad \boldsymbol{u} \neq \boldsymbol{0},$$

so that
$$(\boldsymbol{u}, \boldsymbol{u}) = (T\boldsymbol{u}, T\boldsymbol{u}) = (\lambda\boldsymbol{u}, \lambda\boldsymbol{u}) = \lambda\bar{\lambda}(\boldsymbol{u}, \boldsymbol{u}),$$

and
$$(\lambda\bar{\lambda} - 1)(\boldsymbol{u}, \boldsymbol{u}) = 0.$$

But $(\boldsymbol{u}, \boldsymbol{u})$ is real and positive. Hence $\lambda\bar{\lambda} = 1$, or $|\lambda| = 1$. The real eigenvalues (if any) are therefore equal to either $+1$ or -1.

8.2 Restriction mappings and a normal form for unitary transformations

In order to prove the next theorem we must introduce the concept of *restriction of a mapping*. Let T be a linear mapping of a vector space V and let W be a subspace of V which is mapped into itself by T. Thus $T\boldsymbol{w} \in W$ for all $\boldsymbol{w} \in W$. Let T_W be the mapping of W given by $T_W(\boldsymbol{w}) = T(\boldsymbol{w})$ for all $\boldsymbol{w} \in W$. T_W is clearly a linear mapping and is called the *restriction of* T *to* W. In particular, if T is a unitary mapping on the unitary space V, then T_W is also a unitary mapping on W.

THEOREM 8.3. If V is a unitary space and T a unitary mapping on V, then there exists an orthonormal basis of V consisting entirely of eigenvectors of T.

Proof. Let $e^{i\theta_1}, \ldots, e^{i\theta_p}$ be the *distinct* eigenvalues of T and let W_k be the eigenspace corresponding to the eigenvalue $e^{i\theta_k}$ $(k = 1, \ldots, p)$. We first show that W_j, W_k are orthogonal subspaces of V if $j \neq k$. Let $\boldsymbol{u} \in W_j$, $\boldsymbol{v} \in W_k$. Then

$$T\boldsymbol{u} = e^{i\theta_j}\boldsymbol{u}, \qquad T\boldsymbol{v} = e^{i\theta_k}\boldsymbol{v}$$

and
$$\begin{aligned}(\boldsymbol{u}, \boldsymbol{v}) &= (T\boldsymbol{u}, T\boldsymbol{v}) = (e^{i\theta_j}\boldsymbol{u}, e^{i\theta_k}\boldsymbol{v}) \\ &= e^{i\theta_j}e^{-i\theta_k}(\boldsymbol{u}, \boldsymbol{v}).\end{aligned}$$

Thus
$$(e^{i\theta_j} - e^{i\theta_k})(\boldsymbol{u}, \boldsymbol{v}) = 0.$$

But $e^{i\theta_j} \neq e^{i\theta_k}$, so that $(\boldsymbol{u}, \boldsymbol{v}) = 0$ and $\boldsymbol{u}, \boldsymbol{v}$ are orthogonal. It follows that $W_1, \ldots, W_p$ are orthogonal subspaces of V. Let $W = W_1 + \cdots + W_p$, so that W is a subspace of V. By Theorem 7.1, $V = W \oplus W^{\perp}$. Let $\boldsymbol{u} \in W^{\perp}$ and $\boldsymbol{w}_k \in W_k$. Then

$$0 = (\boldsymbol{w}_k, \boldsymbol{u}) = (T\boldsymbol{w}_k, T\boldsymbol{u}) = (e^{i\theta_k}\boldsymbol{w}_k, T\boldsymbol{u}) = e^{i\theta_k}(\boldsymbol{w}_k, T\boldsymbol{u}).$$

But $e^{i\theta_k} \neq 0$, so that $(\boldsymbol{w}_k, T\boldsymbol{u}) = 0$, $(k = 1, \ldots, p)$.

It follows that $(\boldsymbol{w}, T\boldsymbol{u}) = 0$ for all $\boldsymbol{w} \in W$, and $T\boldsymbol{u} \in W^{\perp}$. Thus T maps $W^{\perp}$ into itself and $T_{W^{\perp}}$, the restriction of T to $W^{\perp}$, is a unitary mapping of $W^{\perp}$. Suppose, if possible, that $\dim W^{\perp} = \nu > 0$. Now an equation of degree ν with complex coefficients has ν complex roots (not necessarily distinct), so that $T_{W^{\perp}}$ certainly possesses an eigenvalue $e^{i\theta}$, say, and a corresponding eigenvector $\boldsymbol{q} \in W^{\perp}$. Thus

$$T_{W^{\perp}}\boldsymbol{q} = e^{i\theta}\boldsymbol{q}.$$

But $T_{W^{\perp}}\boldsymbol{q} = T\boldsymbol{q}$, so that $\boldsymbol{q}$ is also an eigenvector of T corresponding to the eigenvalue $e^{i\theta}$. It follows that $e^{i\theta} = e^{i\theta_k}$ for some k, $1 \leqslant k \leqslant p$, and $\boldsymbol{q} \in W$. Thus $\boldsymbol{q} \in W \cap W^{\perp}$, and $\boldsymbol{q} = \boldsymbol{0}$, giving a contradiction. Hence $\nu = \dim W^{\perp} = 0$, and $W^{\perp}$ consists of the zero vector alone. It follows that $V = W = W_1 \oplus \cdots \oplus W_p$.

Let B_k be a set of vectors forming an orthonormal basis for W_k ($k = 1, \ldots, p$) and consider the set of vectors $B_1, \ldots, B_p$. These form an orthonormal set (since $W_1, \ldots, W_p$ are mutually orthogonal) and they are, moreover, a spanning set for V. They therefore form an orthonormal basis for V and so there must be n of them, where $\dim V = n$. But each member of each set B_k is an eigenvector of T, and we have therefore found a basis of V consisting entirely of eigenvectors of T.

Corollary 1. The matrix of T relative to the above basis is diagonal.

Let $\{\boldsymbol{v}_1, \ldots, \boldsymbol{v}_n\}$ be an orthonormal basis of V consisting of eigenvectors of T. Then $T\boldsymbol{v}_k = \lambda_k\boldsymbol{v}_k$, where λ_k is an eigenvalue of T ($k = 1, \ldots, n$) and the matrix of T relative to this basis is

$$\Lambda = \begin{bmatrix} \lambda_1 & 0 \cdots 0 \\ 0 & \lambda_2 \cdots 0 \\ \cdot & \cdot\ \cdot\ \cdot\ \cdot \\ 0 & 0 \cdots \lambda_n \end{bmatrix}$$

where $|\lambda_k| = 1$ ($k = 1, \ldots, n$).

Corollary 2. If $\boldsymbol{A}$ is a unitary matrix, then there exists a unitary matrix $\boldsymbol{U}$ such that $\overline{\boldsymbol{U}}^t\boldsymbol{A}\boldsymbol{U} = \Lambda$, where Λ is a diagonal matrix whose diagonal elements all have modulus 1.

$\boldsymbol{A}$ is the matrix of a unitary transformation T relative to an orthonormal basis. By Corollary 1, there exists an orthonormal basis with respect to which T has matrix Λ. Let $\boldsymbol{U}$ be the matrix of the mapping that transforms the first basis into the second. Then $\boldsymbol{U}$ is unitary (the reader should verify this) and $\Lambda = \boldsymbol{U}^{-1}\boldsymbol{A}\boldsymbol{U} = \overline{\boldsymbol{U}}^{\mathrm{t}}\boldsymbol{A}\boldsymbol{U}$.

8.3 Complexification of a real vector space

We should now like to obtain the corresponding results for orthogonal in place of unitary matrices. Since a unitary matrix with real elements is an orthogonal matrix, one might expect that the results for orthogonal matrices can be obtained immediately from those for unitary matrices as a special case. This, unfortunately, is not true, for the following reason. In Theorem 8.3 we made use of the fact that every polynomial equation with complex coefficients has at least one complex root and, in particular, that every unitary operator has an eigenvalue. The corresponding result, that every polynomial equation with real coefficients has at least one real root, is not true. The equation $\lambda^2 - \sqrt{2}\lambda + 1 = 0$ has no real root, for example, and the orthogonal transformation with matrix

$$\begin{bmatrix} 1/\sqrt{2} & -1/\sqrt{2} \\ 1/\sqrt{2} & 1/\sqrt{2} \end{bmatrix}$$

which has this as characteristic equation, has no eigenvalue in $\boldsymbol{R}$. The orthogonal transformation is therefore more difficult to deal with than the unitary one, and it was for this reason that we considered unitary transformations first.

To simplify our task, we introduce the idea of *complexification* of a real vector space. Let V be a real vector space of dimension n and let V^* be the set of all elements of the form $\boldsymbol{p} + i\boldsymbol{q}$, where $\boldsymbol{p}, \boldsymbol{q} \in V$, $i^2 = -1$. We define addition, and multiplication by a complex number $x + iy$ by

$$(\boldsymbol{p}_1 + i\boldsymbol{q}_1) + (\boldsymbol{p}_2 + i\boldsymbol{q}_2) = (\boldsymbol{p}_1 + \boldsymbol{p}_2) + i(\boldsymbol{q}_1 + \boldsymbol{q}_2),$$
$$(x + iy)(\boldsymbol{p} + i\boldsymbol{q}) = (x\boldsymbol{p} - y\boldsymbol{q}) + i(x\boldsymbol{q} + y\boldsymbol{p}).$$

The zero vector in V^* is $\boldsymbol{0} + i\boldsymbol{0}$. Clearly, V^* is a vector space over the complex field $\boldsymbol{C}$. It is also a vector space over the real field $\boldsymbol{R}$. If we identify each vector $\boldsymbol{p} \in V$ with the vector $\boldsymbol{p} + i\boldsymbol{0} \in V^*$, we see that V^* includes V.

We now show that V and V^* have the same dimension and that every basis of V is also a basis of V^*. Let $\{\boldsymbol{v}_1, \ldots, \boldsymbol{v}_n\}$ be a basis of V. Then, if $x_1, \ldots, x_n \in \boldsymbol{R}$,

$$x_1\boldsymbol{v}_1 + x_2\boldsymbol{v}_2 + \cdots + x_n\boldsymbol{v}_n = \boldsymbol{0}$$

implies $x_1 = x_2 = \cdots = x_n = 0$. Now suppose that

$$(x_1 + iy_1)\boldsymbol{v}_1 + (x_2 + iy_2)\boldsymbol{v}_2 + \cdots + (x_n + iy_n)\boldsymbol{v}_n = \boldsymbol{0},$$

where $x_1, \ldots, x_n, y_1, \ldots, y_n \in \boldsymbol{R}$. Then

$$(x_1\boldsymbol{v}_1 + x_2\boldsymbol{v}_2 + \cdots + x_n\boldsymbol{v}_n) + i(y_1\boldsymbol{v}_1 + y_2\boldsymbol{v}_2 + \cdots + y_n\boldsymbol{v}_n) = \boldsymbol{0} + i\boldsymbol{0},$$

so that
$$x_1\boldsymbol{v}_1 + x_2\boldsymbol{v}_2 + \cdots + x_n\boldsymbol{v}_n = \boldsymbol{0}$$

and
$$y_1\boldsymbol{v}_1 + y_2\boldsymbol{v}_2 + \cdots + y_n\boldsymbol{v}_n = \boldsymbol{0}.$$

Thus
$$x_1 = x_2 = \ldots = x_n = 0,\ y_1 = y_2 = \ldots = y_n = 0,$$

and hence
$$x_1 + iy_1 = x_2 + iy_2 = \cdots = x_n + iy_n = 0.$$

We have therefore shown that $\boldsymbol{v}_1, \boldsymbol{v}_2, \ldots, \boldsymbol{v}_n$ is a linearly independent set in V^*.

Now let $\boldsymbol{p} + i\boldsymbol{q}$ be any element in V^*. Since $\boldsymbol{p}, \boldsymbol{q} \in V$, we have the unique representations

$$\boldsymbol{p} = \alpha_1\boldsymbol{v}_1 + \alpha_2\boldsymbol{v}_2 + \cdots + \alpha_n\boldsymbol{v}_n,$$
$$\boldsymbol{q} = \beta_1\boldsymbol{v}_1 + \beta_2\boldsymbol{v}_2 + \cdots + \beta_n\boldsymbol{v}_n.$$

Hence
$$\boldsymbol{p} + i\boldsymbol{q} = (\alpha_1 + i\beta_1)\boldsymbol{v}_1 + \cdots + (\alpha_n + i\beta_n)\boldsymbol{v}_n,$$

and we have shown that $\{\boldsymbol{v}_1, \boldsymbol{v}_2, \ldots, \boldsymbol{v}_n\}$ is also a spanning set for V^*. It follows that $\{\boldsymbol{v}_1, \boldsymbol{v}_2, \ldots, \boldsymbol{v}_n\}$ is a basis for V^*, and V^* has dimension n.

Now let $(\boldsymbol{p}, \boldsymbol{q})$ be an inner product defined on V, so that V is a euclidean space, and define a mapping σ of pairs of elements of V^* into $\boldsymbol{C}$ as follows:

$$\sigma(\boldsymbol{p} + i\boldsymbol{q}, \boldsymbol{r} + i\boldsymbol{s}) = (\boldsymbol{p}, \boldsymbol{r}) + (\boldsymbol{q}, \boldsymbol{s}) + i\{(\boldsymbol{q}, \boldsymbol{r}) - (\boldsymbol{p}, \boldsymbol{s})\}.$$

Writing $\boldsymbol{u} = \boldsymbol{p} + i\boldsymbol{q}$, $\boldsymbol{v} = \boldsymbol{r} + i\boldsymbol{s}$, one can easily verify that

$$\sigma(\boldsymbol{v}, \boldsymbol{u}) = \overline{\sigma(\boldsymbol{u}, \boldsymbol{v})} \qquad \text{(the complex conjugate of } \sigma(\boldsymbol{u}, \boldsymbol{v})\text{)},$$
$$\sigma(x\boldsymbol{u}, \boldsymbol{v}) = x\sigma(\boldsymbol{u}, \boldsymbol{v}) \qquad (x \in \boldsymbol{C}),$$
$$\sigma(\boldsymbol{u}_1 + \boldsymbol{u}_2, \boldsymbol{v}) = \sigma(\boldsymbol{u}_1, \boldsymbol{v}) + \sigma(\boldsymbol{u}_2, \boldsymbol{v}),$$
$$\sigma(\boldsymbol{u}, \boldsymbol{u}) = (\boldsymbol{p}, \boldsymbol{p}) + (\boldsymbol{q}, \boldsymbol{q}) \geqslant 0 \text{ for all } \boldsymbol{u},$$
$$\sigma(\boldsymbol{u}, \boldsymbol{u}) = 0$$

implies
$$\boldsymbol{p} = \boldsymbol{q} = \boldsymbol{0}, \qquad \text{i.e., } \boldsymbol{u} = \boldsymbol{0}.$$

Thus $\sigma(\boldsymbol{u}, \boldsymbol{v})$ defines a hermitian inner product on V^*, and V^* is a unitary space. Moreover, $\sigma(\boldsymbol{u}, \boldsymbol{v}) = (\boldsymbol{u}, \boldsymbol{v})$ when $\boldsymbol{u}, \boldsymbol{v} \in V$.

We can, in a natural way, extend every linear transformation T on V to a linear transformation T^* on V^*. Take

$$T^*(\boldsymbol{p} + i\boldsymbol{q}) = T(\boldsymbol{p}) + iT(\boldsymbol{q}).$$

The reader should verify that T^* is, indeed, a linear mapping. Moreover, T^* coincides with T on V, since $\boldsymbol{v} = \boldsymbol{p} + i\boldsymbol{q} \in V$ if and only if $\boldsymbol{q} = \boldsymbol{0}$, $T(\boldsymbol{q}) = 0$. The above equation then reduces to $T^*(\boldsymbol{v}) = T(\boldsymbol{v})$ if $\boldsymbol{v} \in V$.

Now we have already seen that every basis (necessarily real) of V is also a basis of V^*. It follows that *the matrix of T on V relative to such a basis is the same as the matrix of T^* on V^* relative to the same basis.*

8.4 A normal form for orthogonal matrices

Now take T to be an orthogonal mapping on V, so that

$$(T(\boldsymbol{p}), T(\boldsymbol{q})) = (\boldsymbol{p}, \boldsymbol{q}) \text{ for all } \boldsymbol{p}, \boldsymbol{q} \in V.$$

We then have

$$\begin{aligned} \sigma(T^*(\boldsymbol{u}), T^*(\boldsymbol{v})) &= \sigma(T(\boldsymbol{p}) + iT(\boldsymbol{q}), T(\boldsymbol{r}) + iT(\boldsymbol{s})) \\ &= (T(\boldsymbol{p}), T(\boldsymbol{r})) + (T(\boldsymbol{q}), T(\boldsymbol{s})) + i(T(\boldsymbol{q}), T(\boldsymbol{r})) - i(T(\boldsymbol{p}), T(\boldsymbol{s})) \\ &= (\boldsymbol{p}, \boldsymbol{r}) + (\boldsymbol{q}, \boldsymbol{s}) + i(\boldsymbol{q}, \boldsymbol{r}) - i(\boldsymbol{p}, \boldsymbol{s}) \\ &= \sigma(\boldsymbol{u}, \boldsymbol{v}) \text{ for all } \boldsymbol{u}, \boldsymbol{v} \in V^*. \end{aligned}$$

Thus T^* is a unitary mapping on V^* and its eigenvalues all have unit modulus. They are therefore all of the form $\lambda = e^{i\theta}$, the real eigenvalues (if any) being equal to $+1$ or -1. Let $\lambda = e^{i\theta} \neq \pm 1$ be an eigenvalue and $\boldsymbol{u} = \boldsymbol{p} + i\boldsymbol{q}$ be a corresponding eigenvector. Then

$$T^*(\boldsymbol{p} + i\boldsymbol{q}) = e^{i\theta}(\boldsymbol{p} + i\boldsymbol{q});$$

i.e.,

$$\begin{aligned} T(\boldsymbol{p}) + iT(\boldsymbol{q}) &= (\cos\theta + i\sin\theta)(\boldsymbol{p} + i\boldsymbol{q}) \\ &= \cos\theta\boldsymbol{p} - \sin\theta\boldsymbol{q} + i(\sin\theta\boldsymbol{p} + \cos\theta\boldsymbol{q}). \end{aligned}$$

It follows that

$$T(\boldsymbol{p}) = \cos\theta\boldsymbol{p} - \sin\theta\boldsymbol{q}$$

and

$$T(\boldsymbol{q}) = \sin\theta\boldsymbol{p} + \cos\theta\boldsymbol{q},$$

from which we obtain

$$T(\boldsymbol{p}) - iT(\boldsymbol{q}) = (\cos\theta - i\sin\theta)(\boldsymbol{p} - i\boldsymbol{q}),$$

or

$$T(\boldsymbol{p}) + iT(-\boldsymbol{q}) = e^{-i\theta}(\boldsymbol{p} - i\boldsymbol{q}).$$

Thus

$$T^*(\boldsymbol{p} - i\boldsymbol{q}) = e^{-i\theta}(\boldsymbol{p} - i\boldsymbol{q}).$$

We have shown that $e^{-i\theta}$ is also an eigenvalue and $\overline{\boldsymbol{u}}$ is a corresponding eigenvector.

By Theorem 8.3 there exists an orthonormal basis of V^* consisting entirely of eigenvectors of T^*. Let $V_\lambda{}^*$ be the eigenspace corresponding to the eigenvalue $\lambda = e^{i\theta} \neq \pm 1$, and let $\{\boldsymbol{u}_1, \boldsymbol{u}_2, \ldots, \boldsymbol{u}_m\}$ be an orthonormal basis for V_λ^*. Then $\{\overline{\boldsymbol{u}}_1, \overline{\boldsymbol{u}}_2, \ldots, \overline{\boldsymbol{u}}_m\}$ will be an orthonormal basis for $V_{\bar{\lambda}}^*$ (the reader should check this). Moreover, since $\lambda \neq \bar{\lambda}$, the spaces V_λ^* and $V_{\bar{\lambda}}^*$ are orthogonal, so that $\{\boldsymbol{u}_1, \overline{\boldsymbol{u}}_1, \boldsymbol{u}_2, \overline{\boldsymbol{u}}_2, \ldots, \boldsymbol{u}_m, \overline{\boldsymbol{u}}_m\}$ is an orthonormal basis of $V_\lambda^* + V_{\bar{\lambda}}^*$. Let $\boldsymbol{u}_j = \boldsymbol{p}_j + i\boldsymbol{q}_j$ $(j = 1, \ldots, m)$.

Now, if $j \neq k$,

$$0 = \sigma(\boldsymbol{u}_j, \boldsymbol{u}_k) = (\boldsymbol{p}_j, \boldsymbol{p}_k) + (\boldsymbol{q}_j, \boldsymbol{q}_k) + i(\boldsymbol{q}_j, \boldsymbol{p}_k) - i(\boldsymbol{p}_j, \boldsymbol{q}_k)$$

and also

$$0 = \sigma(\boldsymbol{u}_j, \overline{\boldsymbol{u}}_k) = (\boldsymbol{p}_j, \boldsymbol{p}_k) - (\boldsymbol{q}_j, \boldsymbol{q}_k) + i(\boldsymbol{q}_j, \boldsymbol{p}_k) + i(\boldsymbol{p}_j, \boldsymbol{q}_k).$$

This gives

$$(\boldsymbol{p}_j, \boldsymbol{p}_k) + (\boldsymbol{q}_j, \boldsymbol{q}_k) = 0 = (\boldsymbol{p}_j, \boldsymbol{p}_k) - (\boldsymbol{q}_j, \boldsymbol{q}_k)$$

and

$$(\boldsymbol{q}_j, \boldsymbol{p}_k) - (\boldsymbol{p}_j, \boldsymbol{q}_k) = 0 = (\boldsymbol{q}_j, \boldsymbol{p}_k) + (\boldsymbol{p}_j, \boldsymbol{q}_k).$$

Hence

$$(\boldsymbol{p}_j, \boldsymbol{p}_k) = (\boldsymbol{q}_j, \boldsymbol{q}_k) = (\boldsymbol{p}_j, \boldsymbol{q}_k) = 0 \qquad (j \neq k).$$

Finally, from $\sigma(\boldsymbol{u}_j, \boldsymbol{u}_j) = 1$ and $\sigma(\boldsymbol{u}_j, \overline{\boldsymbol{u}}_j) = 0$ we see that

$$(\boldsymbol{p}_j, \boldsymbol{p}_j) = (\boldsymbol{q}_j, \boldsymbol{q}_j) = \tfrac{1}{2} \qquad \text{and} \qquad (\boldsymbol{p}_j, \boldsymbol{q}_j) = 0 \qquad (j = 1, \ldots, m).$$

We have therefore shown that the vectors $\boldsymbol{p}_1, \boldsymbol{q}_1, \boldsymbol{p}_2, \boldsymbol{q}_2, \ldots, \boldsymbol{p}_m, \boldsymbol{q}_m$ are mutually orthogonal and that $\{\sqrt{2}\boldsymbol{p}_1, \sqrt{2}\boldsymbol{q}_1, \sqrt{2}\boldsymbol{p}_2, \sqrt{2}\boldsymbol{q}_2, \ldots, \sqrt{2}\boldsymbol{p}_m, \sqrt{2}\boldsymbol{q}_m\}$ is a real orthonormal basis for $V_\lambda^* + V_{\bar{\lambda}}^*$, which contains $V_\lambda + V_{\bar{\lambda}}$. Moreover, for $i = 1, \ldots, m$,

$$T(\boldsymbol{p}_k) = \cos\theta\,\boldsymbol{p}_k - \sin\theta\,\boldsymbol{q}_k,$$
$$T(\boldsymbol{q}_k) = \sin\theta\,\boldsymbol{p}_k + \cos\theta\,\boldsymbol{q}_k.$$

We can clearly find a real orthonormal basis for V_1^* (if it is not empty). For if $T^*(\boldsymbol{u}) = \boldsymbol{u}$, then $T(\boldsymbol{p}) = \boldsymbol{p}$ and $T(\boldsymbol{q}) = \boldsymbol{q}$, so that $\boldsymbol{p}$ and $\boldsymbol{q}$ are real eigenvectors corresponding to the eigenvalue 1. Similarly we can find a real orthonormal basis for V_{-1}^*.

Let us now combine the orthonormal bases for all the eigenspaces to form a real orthonormal basis of V. If we order the basis so that all eigenvectors corresponding to $+1$ come first, followed by those corresponding to -1, and finally the bases of $V_\lambda^* + V_{\bar\lambda}^*$ in succession, each one being in the order $\{\sqrt{2}\mathbf{p}_1, \sqrt{2}\mathbf{q}_1, \ldots, \sqrt{2}\mathbf{p}_m, \sqrt{2}\mathbf{q}_m\}$, then the matrix of T^* (and hence of T) is of the form

$$\begin{bmatrix} 1 &&&&&&&&&&& \\ & 1 &&&&&&&&&& \\ && \cdot &&&&&&&&& \\ &&& \cdot &&&&&&&& \\ &&&& 1 &&&&&&& \\ &&&&& -1 &&&&&& \\ &&&&&& -1 &&&&& \\ &&&&&&& \cdot &&&& \\ &&&&&&&& -1 &&& \\ &&&&&&&&& \cos\theta_1 & -\sin\theta_1 && \\ &&&&&&&&& \sin\theta_1 & \cos\theta_1 && \\ &&&&&&&&&&& \cdot & \\ &&&&&&&&&&&& \cdot \\ &&&&&&&&&&&&& \cos\theta_r & -\sin\theta_r \\ &&&&&&&&&&&&& \sin\theta_r & \cos\theta_r \end{bmatrix}$$

All unindicated elements are zeros and all the elements $+1$, -1, $\cos\theta_1, \ldots, \cos\theta_r$ are in the leading diagonal. $e^{i\theta_1}, e^{i\theta_2}, \ldots, e^{i\theta_r}$ (not necessarily distinct) are the unreal eigenvalues of T^*. The number of 1's is equal to $m(1)$, the multiplicity of 1 as an eigenvalue of T, and the number of -1's is $m(-1)$, the multiplicity of -1.

In a two-dimensional subspace the matrix $\begin{bmatrix} \cos\theta & -\sin\theta \\ \sin\theta & \cos\theta \end{bmatrix}$ represents a rotation through an angle θ, and the matrix $\begin{bmatrix} -1 & 0 \\ 0 & -1 \end{bmatrix}$ represents a rotation through an angle π. A single -1 represents a reflection and each $+1$, of course, represents an identity on a one-dimensional subspace. We say that T is a rotation if dim V_{-1} is even (i.e., the numbers of -1's is even or zero) and a reflection if dim V_{-1} is odd. But

$$\begin{aligned} \det T &= (+1)^{m(1)}(-1)^{m(-1)}(\cos^2\theta_1 + \sin^2\theta_1) \ldots (\cos^2\theta_r + \sin^2\theta_r) \\ &= (-1)^{m(-1)}, \end{aligned}$$

so that $\det T = +1$ if $m(-1)$ is even or zero and $\det T = -1$ if $m(-1)$ is odd. Thus an orthogonal transformation is a rotation if its determinant is $+1$ and a reflection if its determinant is -1.

Now an orthogonal transformation takes an orthonormal basis into an orthonormal basis, and we say that two such bases have the same *orientation* if the orthogonal mapping that transforms one into the other is a rotation.

Example 2. In $\boldsymbol{R}_3$ the orthonormal set of vectors $\boldsymbol{i}, \boldsymbol{j}, \boldsymbol{k}$ form a right-handed system. The vectors $\boldsymbol{u}_1, \boldsymbol{u}_2, \boldsymbol{u}_3$ have coordinate vectors $\{1/2, -\sqrt{3}/2, 0\}$, $\{0, 0, -1\}$, $\{\sqrt{3}/2, 1/2, 0\}$ respectively relative to the basis $\boldsymbol{i}, \boldsymbol{j}, \boldsymbol{k}$. Show that $\boldsymbol{u}_1, \boldsymbol{u}_2, \boldsymbol{u}_3$ is an orthonormal basis of $\boldsymbol{R}_3$ and that these vectors form a right-handed system.

We see by inspection that $(\boldsymbol{u}_l, \boldsymbol{u}_m) = \begin{cases} 0 & (l \neq m) \\ 1 & (l = m) \end{cases}$, so that $\{\boldsymbol{u}_1, \boldsymbol{u}_2, \boldsymbol{u}_3\}$ is an orthonormal basis for $\boldsymbol{R}_3$. Now the transformation T that maps $\boldsymbol{i}, \boldsymbol{j}, \boldsymbol{k}$ on to $\boldsymbol{u}_1, \boldsymbol{u}_2, \boldsymbol{u}_3$ has matrix

$$T = \begin{bmatrix} 1/2 & -\sqrt{3}/2 & 0 \\ 0 & 0 & -1 \\ \sqrt{3}/2 & 1/2 & 0 \end{bmatrix}$$

relative to the basis $\boldsymbol{i}, \boldsymbol{j}, \boldsymbol{k}$. This is an orthogonal matrix and det $\boldsymbol{T} = +1$, so that T is a rotation. It follows that $\boldsymbol{u}_1, \boldsymbol{u}_2, \boldsymbol{u}_3$ has the same orientation as $\boldsymbol{i}, \boldsymbol{j}, \boldsymbol{k}$ and is also a right-handed system.

8.5 An application to R_3

We now show that a rotation in a real three-dimensional space is a *rotation about an axis*. The matrix of an orthogonal transformation on $\boldsymbol{R}_3$ must take one of the normal forms

$$\begin{bmatrix} 1 & 0 & 0 \\ 0 & 1 & 0 \\ 0 & 0 & 1 \end{bmatrix}, \quad \begin{bmatrix} 1 & 0 & 0 \\ 0 & 1 & 0 \\ 0 & 0 & -1 \end{bmatrix}, \quad \begin{bmatrix} 1 & 0 & 0 \\ 0 & -1 & 0 \\ 0 & 0 & -1 \end{bmatrix},$$

$$\begin{bmatrix} -1 & 0 & 0 \\ 0 & -1 & 0 \\ 0 & 0 & -1 \end{bmatrix}, \quad \begin{bmatrix} 1 & 0 & 0 \\ 0 & \cos\theta & -\sin\theta \\ 0 & \sin\theta & \cos\theta \end{bmatrix}, \quad \begin{bmatrix} -1 & 0 & 0 \\ 0 & \cos\theta & -\sin\theta \\ 0 & \sin\theta & \cos\theta \end{bmatrix}.$$

Excluding the first, which is the identity mapping, the only ones with determinant equal to $+1$ are

$$\begin{bmatrix} 1 & 0 & 0 \\ 0 & -1 & 0 \\ 0 & 0 & -1 \end{bmatrix} = \begin{bmatrix} 1 & 0 & 0 \\ 0 & \cos\pi & -\sin\pi \\ 0 & \sin\pi & \cos\pi \end{bmatrix}$$

and

$$\begin{bmatrix} 1 & 0 & 0 \\ 0 & \cos\theta & -\sin\theta \\ 0 & \sin\theta & \cos\theta \end{bmatrix}.$$

These both represent rotations about an axis given by the first basis vector, through angles π and θ respectively. Thus every rotation in a real three-dimensional space is a rotation about an axis. This result is of fundamental importance in mechanics.

Problems

8.1 Prove that the product of two $n \times n$ orthogonal matrices is orthogonal, and show that all such matrices form a multiplicative group.

8.2 Show that all $n \times n$ unitary matrices form a multiplicative group.

8.3 Find an orthogonal matrix having $[5/13, 12/13, 0]$ as its first row.

8.4 Find a unitary matrix whose first row is a multiple of the vector $[1 + i, 1 - i]$.

8.5 What is the complexification of $\boldsymbol{R}_1$?

8.6 λ is an eigenvalue of the matrix $\boldsymbol{A}$ with real elements. If $\boldsymbol{A}$ is orthogonal, prove that $1/\lambda$ is also an eigenvalue.

8.7 If the real matrix $\boldsymbol{B}$ is skew-symmetric (i.e., $\boldsymbol{B}^t = -\boldsymbol{B}$) prove that the matrix $\boldsymbol{I} + \boldsymbol{B}$ is non-singular and the matrix $\boldsymbol{A} = (\boldsymbol{I} - \boldsymbol{B})(\boldsymbol{I} + \boldsymbol{B})^{-1}$ is orthogonal. Taking $\boldsymbol{B} = \begin{bmatrix} 0 & a \\ -a & 0 \end{bmatrix}$, find $\boldsymbol{A}$ and verify that it is orthogonal.

8.8 Verify that the matrix

$$\boldsymbol{A} = \frac{1}{1 + 2a^2} \begin{bmatrix} 1 & -2a & 2a^2 \\ 2a & 1 - 2a^2 & -2a \\ 2a^2 & 2a & 1 \end{bmatrix}$$

is orthogonal and find a matrix $\boldsymbol{B}$ related to $\boldsymbol{A}$ as in Problem 8.7.

8.9 If the matrix $\boldsymbol{B}$ is skew-hermitian (i.e., $\bar{\boldsymbol{B}}^t = -\boldsymbol{B}$) prove that the matrix $\boldsymbol{A} = (\boldsymbol{I} - \boldsymbol{B})(\boldsymbol{I} + \boldsymbol{B})^{-1}$ is unitary. Find $\boldsymbol{A}$ when

$$\boldsymbol{B} = \begin{bmatrix} 0 & i & 0 \\ i & 0 & 0 \\ 0 & 0 & 0 \end{bmatrix}.$$

8.10 $\{\boldsymbol{v}_1, \boldsymbol{v}_2, \ldots, \boldsymbol{v}_n\}$ is an orthonormal basis of the n-dimensional euclidean space V. The transformation T is defined by

$$T\boldsymbol{v} = \sum_{i=1}^{n} (\boldsymbol{v}, \boldsymbol{v}_i)\boldsymbol{v}_i \text{ for all } \boldsymbol{v} \in V.$$

Prove that T is a linear transformation and that $T^2 = T$. Prove also that $(T\boldsymbol{v}, \boldsymbol{w}) = (\boldsymbol{v}, T\boldsymbol{w})$ for all $\boldsymbol{v}, \boldsymbol{w} \in V$.

8.11 T is a unitary mapping of $\boldsymbol{C}_3$. Show that there is a basis of $\boldsymbol{C}_3$ relative to which the matrices of $T, T^2, \ldots, T^k$ are all diagonal, where k is any positive integer. Show that the matrix of $(I + T + T^2 + \ldots + T^k)/(k + 1)$ is also diagonal.

9 · *Quadratic and Hermitian Forms*

A homogeneous polynomial $\sum_{i,j=1}^{n} a_{ij}x_ix_j$ of degree 2 in the n variables $x_1, x_2, \ldots, x_n$ is called a *quadratic form*. We may always suppose that $a_{ij} = a_{ji}$. For if this is not so, we write $b_{ij} = b_{ji} = (a_{ij} + a_{ji})/2$. We then have $\sum_{i,j=1}^{n} a_{ij}x_ix_j = \sum_{i,j=1}^{n} b_{ij}x_ix_j$, where $b_{ij} = b_{ji}$. Let us now suppose that all the elements a_{ij} are real, and regard $x = \{x_1, x_2, \ldots, x_n\}$ as an element in the vector space V_n over the real field. We can then write the quadratic form as $\boldsymbol{x}^t A\boldsymbol{x}$, where A is a real symmetric $n \times n$ matrix.

If we change the basis by writing $\boldsymbol{x} = \boldsymbol{P}\boldsymbol{y}$, where $\boldsymbol{P}$ is non-singular, then

$$\boldsymbol{x}^t A\boldsymbol{x} = (\boldsymbol{P}\boldsymbol{y})^t A(\boldsymbol{P}\boldsymbol{y}) = \boldsymbol{y}^t\boldsymbol{P}^t A\boldsymbol{P}\boldsymbol{y}.$$

The quadratic form in the variables $y_1, y_2, \ldots, y_n$ has matrix $\boldsymbol{P}^t A\boldsymbol{P}$, which is also symmetric.

9.1 Reduction of a quadratic form to a sum of squares

The reader will be familiar with the process of completing the square. We can use it, for example, to write $x_1^2 - 6x_1x_2 + 11x_2^2$ as a sum of squares as follows:

$$x_1^2 - 6x_1x_2 + 11x_2^2 = (x_1 - 3x_2)^2 + 2x_2^2.$$

On the other hand,

$$\begin{aligned} x_1x_2 + x_2x_3 + x_3x_1 &= \tfrac{1}{4}(x_1 + x_2)^2 - \tfrac{1}{4}(x_1 - x_2)^2 + x_3(x_1 + x_2) \\ &= [\tfrac{1}{2}(x_1 + x_2) + x_3]^2 - \tfrac{1}{4}(x_1 - x_2)^2 - x_3^2. \end{aligned}$$

We can extend this to a quadratic form in n variables. Let us first suppose that $a_{ii} = 0$ for each $i = 1, 2, \ldots, n$, but $a_{hk} \neq 0$ for some $h \neq k$. Write

$$\begin{aligned} x_i &= X_i \qquad (i \neq k), \\ x_k &= X_k + X_h. \end{aligned}$$

Then $$\sum_{i,j=1}^{n} a_{ij}x_ix_j = \sum_{i,j\neq k} a_{ij}X_iX_j + (X_k + X_h)\sum_{j} 2a_{jk}X_j.$$

The coefficient of $X_h^2 = 2a_{hk} \neq 0$.

The transformation $\boldsymbol{x} \to \boldsymbol{X}$ is clearly non-singular (it has determinant $+1$) and hence, by a preliminary change of variables, we can obtain a quadratic form in which at least one 'square' term has a non-zero coefficient. By renumbering the variables, if necessary, we can arrange that $a_{11} \neq 0$. Then

$$\begin{aligned} &\sum_{i,j=1}^{n} a_{ij}x_ix_j = a_{11}x_1^2 + 2x_1(a_{12}x_2 + a_{13}x_3 + \cdots + a_{1n}x_n) + \sum_{i,j\neq 1} a_{ij}x_ix_j \\ &= a_{11}\left(x_1 + \frac{a_{12}}{a_{11}}x_2 + \cdots + \frac{a_{1n}}{a_{11}}x_n\right)^2 \\ &\qquad + (\text{a quadratic form in the } (n-1) \text{ variables } x_2, x_3, \ldots, x_n) \\ &= a_{11}y_1^2 + \sum_{i,j=2}^{n} b_{ij}x_ix_j, \text{ say.} \end{aligned}$$

Repeating the process with the quadratic form $\sum b_{ij}x_ix_j$, and continuing in this way, we see that the process certainly terminates after not more than $(n - 1)$ steps. For we are then left with a quadratic form in the single variable x_n, which must be of the form $\alpha_n x_n^2$. We have shown that, by a transformation of the form

$$\begin{aligned} y_1 &= x_1 + c_{12}x_2 + c_{13}x_3 + \cdots\cdots + c_{1n}x_n, \\ y_2 &= \qquad\;\; x_2 + c_{23}x_3 + \cdots\cdots + c_{2n}x_n, \\ &\cdots\cdots\cdots\cdots\cdots\cdots\cdots\cdots \\ y_{n-1} &= \qquad\qquad\qquad\qquad x_{n-1} + c_{n-1,n}x_n, \\ y_n &= \qquad\qquad\qquad\qquad\qquad\qquad x_n, \end{aligned}$$

we can express the quadratic form as a sum of squares

$$\sum_{i,j=1}^{n} a_{ij}x_ix_j = \alpha_1 y_1^2 + \alpha_2 y_2^2 + \cdots + \alpha_n y_n^2.$$

This transformation is non-singular (with determinant $+1$). It is easy to see that, even if we have to perform preliminary transformations as described above, the resulting transformation is still non-singular. Thus we can write

$$\boldsymbol{x} = \boldsymbol{P}\boldsymbol{y},$$

where $\boldsymbol{x} = \{x_1, x_2, \ldots, x_n\}$, $\boldsymbol{y} = \{y_1, y_2, \ldots, y_n\}$ and $\boldsymbol{P}$ is a non-singular $n \times n$ matrix. Again it is clear that if $\alpha_k = 0$, then $\alpha_{k+1} = \alpha_{k+2} = \cdots = \alpha_n = 0$. Hence

$$\sum_{i,j=1}^{n} a_{ij}x_ix_j = \boldsymbol{x}^{\mathrm{t}}\boldsymbol{A}\boldsymbol{x} = \boldsymbol{y}^{\mathrm{t}}\boldsymbol{D}\boldsymbol{y},$$

where $\boldsymbol{D}$ is the diagonal matrix

$$\begin{bmatrix} \alpha_1 & & & & & & & \\ & \alpha_2 & & & & & \text{O} & \\ & & \cdot & & & & & \\ & & & \cdot & & & & \\ & & & & \alpha_r & & & \\ & & & & & 0 & & \\ & \text{O} & & & & & \cdot & \\ & & & & & & & \cdot \\ & & & & & & & & 0 \end{bmatrix}$$

and $\alpha_1, \alpha_2, \ldots, \alpha_r$ are all non-zero, $r \leqslant n$.

Now

$$\boldsymbol{x}^{\mathrm{t}}\boldsymbol{A}\boldsymbol{x} = (\boldsymbol{P}\boldsymbol{y})^{\mathrm{t}}\boldsymbol{A}(\boldsymbol{P}\boldsymbol{y}) = \boldsymbol{y}^{\mathrm{t}}\boldsymbol{P}^{\mathrm{t}}\boldsymbol{A}\boldsymbol{P}\boldsymbol{y} = \boldsymbol{y}^{\mathrm{t}}\boldsymbol{D}\boldsymbol{y}.$$

This must hold for all values of the variables $\boldsymbol{y}$, and hence

$$\boldsymbol{P}^{\mathrm{t}}\boldsymbol{A}\boldsymbol{P} = \boldsymbol{D}.$$

Now $\boldsymbol{P}$ and $\boldsymbol{P}^{\mathrm{t}}$ are non-singular, so that $\boldsymbol{D}$ and $\boldsymbol{A}$ have the same rank, equal to r (see Theorem 4.8).

The above argument also shows that if, by any non-singular change of variables whatever, the quadratic form $\boldsymbol{x}^{\mathrm{t}}\boldsymbol{A}\boldsymbol{x}$ is reduced to a sum of squares, then the number of squares obtained is always equal to the rank of $\boldsymbol{A}$, and is therefore invariant. It is called the *rank of the quadratic form*. The above result can be expressed as a matrix theorem as follows.

THEOREM 9.1. Let $\boldsymbol{A}$ be a real symmetric $n \times n$ matrix. Then there exists a real non-singular matrix $\boldsymbol{P}$ such that the matrix $\boldsymbol{P}^{\mathrm{t}}\boldsymbol{A}\boldsymbol{P}$ is diagonal.

We can simplify the form still further. Suppose the variables $y_1, y_2, \ldots, y_n$ are ordered in such a way that the non-zero scalars $\alpha_1, \alpha_2, \ldots, \alpha_r$ are arranged with all the positive elements $\alpha_1, \ldots, \alpha_s$ first, followed by all the negative ones $\alpha_{s+1}, \ldots, \alpha_r$. We now make a further non-singular transformation of variables

$$\begin{aligned} z_1 &= +\sqrt{\alpha_1} \cdot y_1, \ldots, z_s = +\sqrt{\alpha_s} \cdot y_s, \\ z_{s+1} &= +\sqrt{(-\alpha_{s+1})} \cdot y_{s+1}, \ldots, z_r = +\sqrt{(-\alpha_r)} \cdot y_r, \\ z_{r+1} &= y_{r+1}, \ldots, z_n = y_n. \end{aligned}$$

Then $$\begin{aligned} \alpha_1 y_1^2 + \cdots + \alpha_r y_r^2 &= z_1^2 + \cdots + z_s^2 - z_{s+1}^2 - \cdots - z_r^2 \\ &= \boldsymbol{z}^{\mathrm{t}}\boldsymbol{D}\boldsymbol{z}, \end{aligned}$$

where $\boldsymbol{D}$ is the diagonal matrix having $+1$ in the first s diagonal places, -1 in the next $r - s$ places, and 0 in the final $n - r$ places.

We next show that the number s depends only on the given form, and is independent of the method of reduction. For suppose, if possible, that

$$\begin{aligned} \sum_{i,j=1}^{n} a_{ij}x_i x_j &= z_1^2 + \cdots + z_s^2 - z_{s+1}^2 - \cdots - z_r^2 \\ &= w_1^2 + \cdots + w_t^2 - w_{t+1}^2 - \cdots - w_r^2, \end{aligned}$$

where $s < t$. Consider the equations

$$z_1 = \cdots = z_s = 0 = w_{t+1} = \cdots = w_r = w_{r+1} = \cdots = w_n.$$

There are $n - t + s < n$ equations in the n unknowns $x_1, x_2, \ldots, x_n$ and these equations have a non-trivial solution $\boldsymbol{x}$ (see Theorem 5.1). If $\boldsymbol{x} \neq \boldsymbol{0}$, then $\boldsymbol{w} \neq \boldsymbol{0}$, so that $w_k \neq 0$ for some k, $1 \leqslant k \leqslant t$. For this solution

$$z_1^2 + \cdots + z_s^2 - z_{s+1}^2 - \cdots - z_r^2 \leqslant 0,$$

and $$w_1^2 + \cdots + w_t^2 - w_{t+1}^2 - \cdots - w_r^2 > 0.$$

This is impossible, and hence $s \nless t$. Similarly $t \nless s$, and we have proved that $s = t$. This is *Sylvester's law of inertia*.

THEOREM 9.2. If a quadratic form is reduced to a sum of squares in any manner whatsoever, then the number p of squares with positive coefficients and the number q of squares with negative coefficients are both invariant.

We know that $p + q = r$, the rank of the form. We define $p - q = s$ to be the *signature* of the form, and we also call it the signature of the matrix of the form. Thus a quadratic form is specified completely by its rank and signature.

9.2 Positive definite forms

The quadratic form $\boldsymbol{x}^t\boldsymbol{A}\boldsymbol{x}$ is said to be *positive definite* if $\boldsymbol{x}^t\boldsymbol{A}\boldsymbol{x} > 0$ for all $\boldsymbol{x} \neq \boldsymbol{0}$. We now show that in this case $q = 0$.

There exists a non-singular linear transformation of variables $\boldsymbol{x} = \boldsymbol{P}\boldsymbol{y}$ such that

$$\boldsymbol{x}^t\boldsymbol{A}\boldsymbol{x} = y_1^2 + \cdots + y_p^2 - y_{p+1}^2 - \cdots - y_{p+q}^2.$$

Suppose that $p + q = r < n$ and take $y_n = 1$, $y_k = 0$ $(k \neq n)$. Then $\boldsymbol{x}^t\boldsymbol{A}\boldsymbol{x} = 0$, but $\boldsymbol{x} \neq \boldsymbol{0}$, since $\boldsymbol{x} = \boldsymbol{0}$ implies $\boldsymbol{y} = \boldsymbol{P}^{-1}\boldsymbol{x} = \boldsymbol{0}$. Thus, if $\boldsymbol{x}^t\boldsymbol{A}\boldsymbol{x}$ is positive definite, we must have $r = n$, and then

$$\boldsymbol{x}^t\boldsymbol{A}\boldsymbol{x} = y_1^2 + \cdots + y_p^2 - y_{p+1}^2 - \cdots - y_n^2.$$

Now suppose that $p < n$ and take $y_{p+1} = 1$, $y_k = 0$ $(k \neq p + 1)$. Then $\boldsymbol{x}^t\boldsymbol{A}\boldsymbol{x} = -y_{p+1}^2 < 0$, again giving a contradiction. It follows that $p = n$.

Conversely, if $p = r = n$,

$$\boldsymbol{x}^t\boldsymbol{A}\boldsymbol{x} = y_1^2 + \cdots + y_n^2$$

and the form is obviously positive definite. Thus the condition $p = r = n$ is necessary and sufficient for a quadratic form to be positive definite. In this case $\boldsymbol{P}^t\boldsymbol{A}\boldsymbol{P} = \boldsymbol{I}$, the unit matrix, for some non-singular matrix $\boldsymbol{P}$.

9.3 Hermitian forms

The results obtained for quadratic forms can easily be generalized by replacing the real field by the complex field. A polynomial of the form $\sum_{i,j=1}^{n} a_{ij}\bar{x}_i x_j$, where $\bar{a}_{ij} = a_{ji}$, is called a *hermitian form*. If $\boldsymbol{x} = \{x_1, \ldots, x_n\}$ is an element of the vector space V_n over the complex field, we can write the form as a matrix product

$$\bar{\boldsymbol{x}}^t\boldsymbol{A}\boldsymbol{x},$$

where $\bar{\boldsymbol{A}}^t = \boldsymbol{A}$, so that $\boldsymbol{A}$ is a hermitian matrix. If we change the basis of V_n we make a transformation of the form $\boldsymbol{x} = \boldsymbol{Py}$, where $\boldsymbol{P}$ is non-singular. We then have

$$\bar{\boldsymbol{x}}^{\mathrm{t}}\boldsymbol{Ax} = \overline{(\boldsymbol{Py})}^{\mathrm{t}}\boldsymbol{A}(\boldsymbol{Py}) = \bar{\boldsymbol{y}}^{\mathrm{t}}\bar{\boldsymbol{P}}^{\mathrm{t}}\boldsymbol{APy} = \bar{\boldsymbol{y}}^{\mathrm{t}}\boldsymbol{By},$$

where $\boldsymbol{B} = \bar{\boldsymbol{P}}^{\mathrm{t}}\boldsymbol{AP}$. Now

$$\bar{\boldsymbol{B}}^{\mathrm{t}} = \overline{(\bar{\boldsymbol{P}}^{\mathrm{t}}\boldsymbol{AP})}^{\mathrm{t}} = \bar{\boldsymbol{P}}^{\mathrm{t}}\bar{\boldsymbol{A}}^{\mathrm{t}}\boldsymbol{P} = \bar{\boldsymbol{P}}^{\mathrm{t}}\boldsymbol{AP} = \boldsymbol{B},$$

so that $\boldsymbol{B}$ is also hermitian. Hence, relative to any basis, the matrix of a hermitian form is hermitian. Moreover, the rank of $\boldsymbol{B}$ is equal to the rank r of $\boldsymbol{A}$, and we define this to be the rank of the hermitian form. Again we can show that, by a non-singular change of variables, we can write

$$\sum_{i,j=1}^{n} a_{ij}\bar{x}_i x_j = \alpha_1|y_1|^2 + \cdots + \alpha_r|y_r|^2,$$

or

$$\bar{\boldsymbol{x}}^{\mathrm{t}}\boldsymbol{Ax} = \bar{\boldsymbol{y}}^{\mathrm{t}}\boldsymbol{Dy},$$

where $\boldsymbol{D}$ is the diagonal matrix

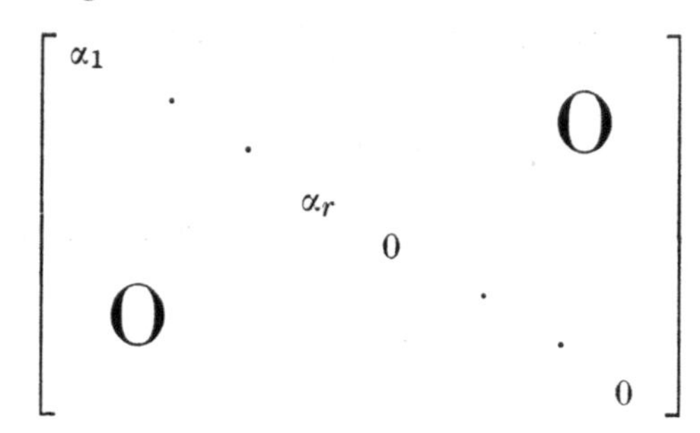

But $\boldsymbol{D}$ is hermitian, so that $\bar{\alpha}_k = \alpha$ for each k, and hence $\alpha_1, \ldots, \alpha_r$ are all real. It follows that the hermitian form $\bar{\boldsymbol{x}}^{\mathrm{t}}\boldsymbol{Ax}$ is real for all values of $\boldsymbol{x}$. We can again establish that the number of positive α's is invariant, and define the signature of the form as before. Since a hermitian form is real we can define a positive definite form as before, and it is easy to see that the condition $p = r = n$ is necessary and sufficient for a form to be positive definite. In this case $\bar{\boldsymbol{P}}^{\mathrm{t}}\boldsymbol{AP} = \boldsymbol{I}$ for some non-singular matrix $\boldsymbol{P}$.

9.4 Eigenvalues of hermitian and symmetric matrices

THEOREM 9.3. The eigenvalues of a hermitian matrix are real.

Proof. Let $\boldsymbol{A}$ be hermitian and $\boldsymbol{Ax} = \lambda\boldsymbol{x}$, $\boldsymbol{x} \neq \boldsymbol{0}$. Then

$$\bar{\boldsymbol{x}}^{\mathrm{t}}\boldsymbol{Ax} = \lambda\bar{\boldsymbol{x}}^{\mathrm{t}}\boldsymbol{x}.$$

Now $\bar{x}^t A x$ and $\bar{x}^t x$ are both hermitian forms and therefore real, and $x^t x \neq 0$ since $x \neq 0$. Hence λ is real and finite.

Corollary. The eigenvalues of a real symmetric matrix are real.

This follows at once from the fact that a real symmetric matrix is a special case of a hermitian matrix. This implies that the characteristic equation of a real symmetric matrix has n real (not necessarily distinct) roots.

9.5 Self-adjoint transformations

We now consider an important class of linear transformations on an inner product space. The linear mapping T on the inner product space V is said to be *self-adjoint* if

$$(Tx, y) = (x, Ty)$$

for all vectors $x, y \in V$.

If V is the n-dimensional space $V_n(C)$ with the usual inner product defined by

$$(x, y) = x^t \bar{y},$$

and if T is the matrix of the transformation T relative to a given orthonormal basis, then

$$(Tx)^t \bar{y} = x^t \overline{(Ty)},$$

or

$$x^t T^t \bar{y} = x^t \overline{Ty} \text{ for all } x, y \in V_n(C).$$

Thus $T^t = \bar{T}$, or $\bar{T}^t = T$, and the matrix T is hermitian.

This argument can be reversed, so that *a linear operator on a unitary space is self-adjoint if and only if its matrix relative to an orthonormal basis is hermitian.*

By Theorem 9.3, it follows that the eigenvalues of a self-adjoint linear mapping are real. We now prove the following theorem.

THEOREM 9.4. Let T be a self-adjoint linear mapping of the unitary space $V_n(C)$. Then there exists an orthonormal basis of V_n consisting entirely of eigenvectors of T.

Proof. Let $\lambda_1, \lambda_2, \ldots, \lambda_k$ be the *distinct* eigenvalues of T. We have seen that they are necessarily real. Let E_h be the eigenspace corresponding to the eigenvalue λ_h, $h = 1, 2, \ldots, k$. We first show that the subspaces E_p and E_q are orthogonal if $p \neq q$. Let $\boldsymbol{x} \in E_p$ and $\boldsymbol{y} \in E_q$. Then

$$T\boldsymbol{x} = \lambda_p \boldsymbol{x} \qquad \text{and} \qquad T\boldsymbol{y} = \lambda_q \boldsymbol{y}.$$

Hence

$$\begin{aligned} \lambda_p(\boldsymbol{x}, \boldsymbol{y}) &= (\lambda_p \boldsymbol{x}, \boldsymbol{y}) = (T\boldsymbol{x}, \boldsymbol{y}) \\ &= (\boldsymbol{x}, T\boldsymbol{y}) = (\boldsymbol{x}, \lambda_q \boldsymbol{y}) = \lambda_q(\boldsymbol{x}, \boldsymbol{y}). \end{aligned}$$

Since $\lambda_p \neq \lambda_q, \qquad (\boldsymbol{x}, \boldsymbol{y}) = 0 \qquad \text{and} \qquad E_p \perp E_q.$

Now let

$$E = E_1 \oplus E_2 \oplus \cdots \oplus E_k.$$

Any vector $\boldsymbol{v} \in E$ can be expressed in the form

$$\boldsymbol{v} = \boldsymbol{v}_1 + \boldsymbol{v}_2 + \cdots + \boldsymbol{v}_k,$$

where $\boldsymbol{v}_h \in E_h$ $(h = 1, \ldots, k)$.

Then

$$\begin{aligned} T\boldsymbol{v} &= T\boldsymbol{v}_1 + T\boldsymbol{v}_2 + \cdots + T\boldsymbol{v}_k \\ &= \lambda_1\boldsymbol{v}_1 + \lambda_2\boldsymbol{v}_2 + \cdots + \lambda_k\boldsymbol{v}_k \in E. \end{aligned}$$

Let $\boldsymbol{w} \in E^\perp$. Then $(\boldsymbol{w}, \boldsymbol{v}) = 0$ for all $\boldsymbol{v} \in E$. Hence $(\boldsymbol{w}, T\boldsymbol{v}) = 0$, so that $(T\boldsymbol{w}, \boldsymbol{v}) = (\boldsymbol{w}, T\boldsymbol{v}) = 0$ for all $\boldsymbol{v} \in E$. It follows that $T\boldsymbol{w} \in E^\perp$ whenever $\boldsymbol{w} \in E^\perp$. Let T_e denote the restriction of T to $E^\perp$. If dim $E^\perp > 0$, then T_e has at least one eigenvalue λ, since the characteristic equation of T_e is of degree at least one, and has at least one root in the complex field. Hence there exists a non-zero vector $\boldsymbol{w} \in E^\perp$ such that $T_e\boldsymbol{w} = \lambda\boldsymbol{w}$. But T_e coincides with T on $E^\perp$, and so λ must also be an eigenvalue of T, and so must coincide with one of the λ_h. But this means that $\boldsymbol{w} \in E_h$, i.e., $\boldsymbol{w} \in E$, which contradicts the fact that $V = E \oplus E^\perp$. Hence dim $E^\perp \ngtr 0$, so that dim $E^\perp = 0$, and $E^\perp$ consists of the zero vector alone. We have just shown that

$$V_n = E_1 \oplus E_2 \oplus \cdots \oplus E_k.$$

If we now take an orthonormal basis for each of the subspaces E_h $(h = 1, \ldots, k)$ consisting, of course, entirely of eigenvectors of T, and combine these, we obtain an orthonormal basis for V_n consisting entirely of eigenvectors of T.

Corollary 1. Let $\boldsymbol{A}$ be an $n \times n$ hermitian matrix. Then there exists an $n \times n$ unitary matrix $\boldsymbol{P}$ such that

$$\boldsymbol{P}^{-1}\boldsymbol{A}\boldsymbol{P} = \overline{\boldsymbol{P}}^{t}\boldsymbol{A}\boldsymbol{P} = \Lambda,$$

Λ being a diagonal matrix.

Proof. $\boldsymbol{A}$ is the matrix of a self-adjoint mapping of V_n, relative to an orthonormal basis. By the theorem, there exists a basis of V_n consisting of eigenvectors $\boldsymbol{v}_1, \boldsymbol{v}_2, \ldots, \boldsymbol{v}_n$ of $\boldsymbol{A}$. Thus

$$\boldsymbol{A}\boldsymbol{v}_h = \lambda_h \boldsymbol{v}_h \qquad (h = 1, \ldots, n), \tag{1}$$

where $\lambda_1, \ldots, \lambda_n$ are the eigenvalues of $\boldsymbol{A}$, not necessarily distinct, but certainly real. We can certainly choose the $\boldsymbol{v}_h$ so that $(\boldsymbol{v}_h, \boldsymbol{v}_h) = 1$ $(h = 1, \ldots, n)$. Let $\boldsymbol{P}$ be the $n \times n$ matrix having $\boldsymbol{v}_1, \boldsymbol{v}_2, \ldots, \boldsymbol{v}_n$ as its columns. The columns form an orthonormal set, so that $\boldsymbol{P}$ is a unitary matrix, and the n equations (1) can be written in the form

$$\boldsymbol{AP} = \boldsymbol{P}\Lambda,$$

where Λ is the diagonal matrix

$$\begin{bmatrix} \lambda_1 & & & & \\ & \lambda_2 & & & \mathrm{O} \\ & & \cdot & & \\ & \mathrm{O} & & \cdot & \\ & & & & \cdot \\ & & & & \lambda_n \end{bmatrix}.$$

Hence $$\boldsymbol{P}^{-1}\boldsymbol{AP} = \Lambda,$$

or $$\bar{\boldsymbol{P}}^{\mathrm{t}}\boldsymbol{AP} = \Lambda \qquad (\text{since } \boldsymbol{P}^{-1} = \bar{\boldsymbol{P}}^{\mathrm{t}}).$$

Corollary 2. If $\bar{\boldsymbol{x}}^{\mathrm{t}}\boldsymbol{A}\boldsymbol{x}$ is a hermitian form, there exists a unitary transformation which transforms it into a sum of squares.

Proof. There exists a unitary matrix $\boldsymbol{P}$ such that $\bar{\boldsymbol{P}}^{\mathrm{t}}\boldsymbol{AP} = \Lambda$. Write $\boldsymbol{x} = \boldsymbol{P}\boldsymbol{y}$. Then

$$\begin{aligned} \bar{\boldsymbol{x}}^{\mathrm{t}}\boldsymbol{A}\boldsymbol{x} &= \bar{\boldsymbol{y}}^{\mathrm{t}}\bar{\boldsymbol{P}}^{\mathrm{t}}\boldsymbol{AP}\boldsymbol{y} = \bar{\boldsymbol{y}}^{\mathrm{t}}\Lambda\boldsymbol{y} \\ &= \lambda_1|y_1|^2 + \lambda_2|y_2|^2 + \cdots + \lambda_n|y_n|^2. \end{aligned}$$

We now give the corresponding results for a euclidean (in place of a unitary) space. Let T be a self-adjoint linear mapping of the euclidean space $V_n(\boldsymbol{R})$. Then the matrix of T relative to any orthonormal basis is real and symmetric. Since a real symmetric matrix is certainly hermitian, its eigenvalues are real and its eigenvectors are real also. As in Theorem 9.4, we can show that there exists an orthonormal basis of $V_n(\boldsymbol{R})$ consisting entirely of eigenvectors of T.

We can then prove results analogous to Theorem 9.4, Corollaries 1 and 2. The matrix $\boldsymbol{P}$ in Corollary 1 now has *real* columns forming an orthonormal set, so that $\boldsymbol{P}$ is orthogonal in this case, and $\boldsymbol{P}^{-1} = \boldsymbol{P}^{\mathrm{t}}$. We obtain the following theorem.

THEOREM 9.5. Let T be a self-adjoint linear mapping of the euclidean space $V_n(\boldsymbol{R})$. Then there exists an orthonormal basis of V_n consisting entirely of eigenvectors of T.

Corollary 1. Let $\boldsymbol{A}$ be an $n \times n$ real symmetric matrix. Then there exists an $n \times n$ orthogonal matrix $\boldsymbol{P}$ such that

$$\boldsymbol{P}^{-1}\boldsymbol{A}\boldsymbol{P} = \boldsymbol{P}^{\mathrm{t}}\boldsymbol{A}\boldsymbol{P} = \Lambda,$$

Λ being a diagonal matrix.

Corollary 2. If $\boldsymbol{x}^{\mathrm{t}}\boldsymbol{A}\boldsymbol{x}$ is a quadratic form, $\boldsymbol{A}$ being a real symmetric matrix, there exists an orthogonal transformation which transforms it into a sum of squares.

Proof. There exists an orthogonal matrix $\boldsymbol{P}$ such that $\boldsymbol{P}^{\mathrm{t}}\boldsymbol{A}\boldsymbol{P} = \Lambda$. Write $\boldsymbol{x} = \boldsymbol{P}\boldsymbol{y}$, so that $\boldsymbol{y}$ is real. Then

$$\begin{aligned}\boldsymbol{x}^{\mathrm{t}}\boldsymbol{A}\boldsymbol{x} &= \boldsymbol{y}^{\mathrm{t}}\boldsymbol{P}^{\mathrm{t}}\boldsymbol{A}\boldsymbol{P}\boldsymbol{y} = \boldsymbol{y}^{\mathrm{t}}\Lambda\boldsymbol{y} \\ &= \lambda_1 y_1^2 + \lambda_2 y_2^2 + \cdots + \lambda_n y_n^2.\end{aligned}$$

9.6 A necessary and sufficient condition for a form to be positive definite

THEOREM 9.6. The quadratic form $\boldsymbol{x}^{\mathrm{t}}\boldsymbol{A}\boldsymbol{x}$ in n variables is positive definite if and only if the determinants

$$|a_{11}|, \quad \begin{vmatrix} a_{11} & a_{12} \\ a_{21} & a_{22} \end{vmatrix}, \quad \begin{vmatrix} a_{11} & a_{12} & a_{13} \\ a_{21} & a_{22} & a_{23} \\ a_{31} & a_{32} & a_{33} \end{vmatrix}, \ldots, \quad \begin{vmatrix} a_{11} & \cdots & a_{1n} \\ \cdot & \cdots & \cdot \\ \cdot & \cdots & \cdot \\ a_{n1} & & a_{nn} \end{vmatrix}$$

are all positive.

Proof. As in Section 9.1, we can express the quadratic form as a sum of squares

$$\boldsymbol{x}^{\mathrm{t}}\boldsymbol{A}\boldsymbol{x} = \sum_{i,j=1}^{n} a_{ij}x_i x_j = \sum_{k=1}^{n} \alpha_k y_k^2 = \boldsymbol{y}^{\mathrm{t}}\boldsymbol{D}\boldsymbol{y},$$

where $\boldsymbol{x} = \boldsymbol{P}\boldsymbol{y}$ and the matrix $\boldsymbol{P}$ has determinant 1. Thus $\boldsymbol{P}^{\mathrm{t}}\boldsymbol{A}\boldsymbol{P} = \boldsymbol{D}$. Suppose $\boldsymbol{x}^{\mathrm{t}}\boldsymbol{A}\boldsymbol{x}$ to be positive definite. Then $\boldsymbol{y}^{\mathrm{t}}\boldsymbol{D}\boldsymbol{y}$ is also positive definite, since $\boldsymbol{x} = \boldsymbol{0}$ if and only if $\boldsymbol{y} = \boldsymbol{0}$. Hence $\alpha_k > 0$ for each k, and $\det \boldsymbol{D} = \alpha_1\alpha_2 \ldots \alpha_n > 0$. But $\det \boldsymbol{P} = \det \boldsymbol{P}^{\mathrm{t}} = 1$, and it follows that $\det \boldsymbol{D} = \det \boldsymbol{A} > 0$.

Let $\boldsymbol{A}_{(k)}$ denote the matrix formed from $\boldsymbol{A}$ by deleting all but its first k rows and all but its first k columns, and let $\boldsymbol{x}_{(k)}$ denote the vector

$\{x_1, x_2, \ldots, x_k\}$. It is easily seen that $\boldsymbol{x}_{(k)} = \boldsymbol{P}_{(k)}\boldsymbol{y}_{(k)}$, where det $\boldsymbol{P}_{(k)} =$ 1. Now take

$$x_{k+1} = x_{k+2} = \cdots = x_n = 0,$$

so that

$$y_{k+1} = y_{k+2} = \cdots = y_n = 0.$$

We then have

$$\sum_{i,j=1}^{k} a_{ij}x_ix_j = \sum_{i=1}^{k} \alpha_i y_i^2,$$

i.e.,

$$\boldsymbol{x}_{(k)}^t \boldsymbol{A}_{(k)} \boldsymbol{x}_{(k)} = \boldsymbol{y}_{(k)}^t \boldsymbol{D}_{(k)} \boldsymbol{y}_{(k)},$$

where, as before,

$$\det \boldsymbol{A}_{(k)} = \det \boldsymbol{D}_{(k)} = \alpha_1\alpha_2 \ldots \alpha_k > 0.$$

This holds for $k = 1, 2, \ldots, n$, so that the conditions are necessary.

Conversely, suppose that det $\boldsymbol{A}_{(k)} > 0$ $(k = 1, 2, \ldots, n)$. Since det $\boldsymbol{A}_{(k)} = \alpha_1\alpha_2 \ldots \alpha_k$, we have

$$\alpha_1 > 0, \qquad \alpha_1\alpha_2 > 0, \qquad \alpha_1\alpha_2\alpha_3 > 0, \qquad \ldots, \alpha_1\alpha_2 \ldots \quad \alpha_n > 0.$$

Hence $\alpha_1, \alpha_2, \ldots, \alpha_n$ are all positive, and

$$\boldsymbol{x}^t \boldsymbol{A} \boldsymbol{x} = \sum_{i=1}^{n} a_i y^2$$

is positive definite. The conditions are therefore sufficient.

If $\overline{\boldsymbol{x}}^t \boldsymbol{A} \boldsymbol{x}$ (where $\overline{\boldsymbol{A}}^t = \boldsymbol{A}$) is a hermitian form, then

$$\overline{\boldsymbol{A}}_{(k)} = \boldsymbol{A}_{(k)}^t \qquad (1 \leqslant k \leqslant n),$$

so that

$$\overline{\det \boldsymbol{A}_{(k)}} = \det \boldsymbol{A}_{(k)}^t = \det \boldsymbol{A}_{(k)}.$$

Hence det $\boldsymbol{A}_{(k)}$ is real. Exactly as above we can now show that the hermitian form $\overline{\boldsymbol{x}}^t \boldsymbol{A} \boldsymbol{x}$ is positive definite if and only if det $\boldsymbol{A}_{(k)} > 0$ $(1 \leqslant k \leqslant n)$.

9.7 Applications

The equation of a central conic with centre O, referred to rectangular cartesian axes Ox_1, Ox_2, is of the form

$$ax_1^2 + 2hx_1x_2 + bx_2^2 = 1,$$

or

$$\boldsymbol{x}^t \boldsymbol{A} \boldsymbol{x} = 1,$$

where $\boldsymbol{x}$ is the column vector $\{x_1, x_2\}$ and $\boldsymbol{A} = \begin{bmatrix} a & h \\ h & b \end{bmatrix}$.

Since $\boldsymbol{A}$ is real and symmetric we can find an orthogonal matrix $\boldsymbol{P}$ such that $\boldsymbol{P}^{\mathrm{t}}\boldsymbol{A}\boldsymbol{P} = \Lambda$, where Λ is a diagonal matrix. We can regard $\boldsymbol{x}$ as an element of the euclidean space $V_2(\boldsymbol{R})$, and change to a new orthonormal basis by means of the orthogonal transformation with matrix $\boldsymbol{P}$. This is equivalent to a rotation of the axes. Thus the coordinates (y_1, y_2) relative to new rectangular axes Oy_1, Oy_2 are given by $\boldsymbol{x} = \boldsymbol{P}\boldsymbol{y}$. The equation of the conic becomes

$$(\boldsymbol{P}\boldsymbol{y})^{\mathrm{t}}\boldsymbol{A}(\boldsymbol{P}\boldsymbol{y}) = 1,$$

i.e.,

$$\boldsymbol{y}^{\mathrm{t}}\boldsymbol{P}^{\mathrm{t}}\boldsymbol{A}\boldsymbol{P}\boldsymbol{y} = 1,$$

or

$$\boldsymbol{y}^{\mathrm{t}}\Lambda\boldsymbol{y} = 1.$$

This equation, which can be written

$$\lambda_1 y_1^2 + \lambda_2 y_2^2 = 1,$$

is the equation of a conic referred to its principal axes as axes of reference. If λ_1 and λ_2 are both positive, the conic is an ellipse with axes of length $2/(\sqrt{\lambda_1})$, $2/(\sqrt{\lambda_2})$. If λ_1 and λ_2 have opposite signs, the conic is a hyperbola with axes of length $2/(\sqrt{|\lambda_1|})$, $2/(\sqrt{|\lambda_2|})$. If λ_1 and λ_2 are both negative, there are no real points on the conic.

Example 1. Find the nature and the equations and lengths of the axes of the conics

(a) $5x_1^2 - 6x_1x_2 + 5x_2^2 = 1$,
(b) $x_1^2 + 12x_1x_2 + 6x_2^2 = 4$.

(a) The conic has equation $\boldsymbol{x}^{\mathrm{t}}\boldsymbol{A}\boldsymbol{x} = 1$, where

$$\boldsymbol{A} = \begin{bmatrix} 5 & -3 \\ -3 & 5 \end{bmatrix}.$$

The eigenvalues are given by

$$\begin{bmatrix} 5-\lambda & -3 \\ -3 & 5-\lambda \end{bmatrix} = 0,$$

$$(5-\lambda)^2 - 9 = 0,$$

$$\lambda = 2 \text{ or } 8.$$

For $\lambda = 2$, the corresponding eigenvector is given by

$$\begin{bmatrix} 3 & -3 \\ -3 & 3 \end{bmatrix}\begin{bmatrix} x_1 \\ x_2 \end{bmatrix} = \boldsymbol{0}.$$

Thus $x_1 = x_2$ and we may take $\boldsymbol{x} = \{1, 1\}$. We normalize this by dividing by $\sqrt{(1^2 + 1^2)} = \sqrt{2}$, giving $\boldsymbol{x} = \{1/\sqrt{2}, 1/\sqrt{2}\}$. Similarly when $\lambda = 8$, $x_1 + x_2 = 0$, $\boldsymbol{x} = \{1/\sqrt{2}, -1/\sqrt{2}\}$.

As we should expect, since the eigenvalues are distinct, these eigenvectors are orthogonal and the 2×2 matrix $\boldsymbol{P}$ having these as its columns is an orthogonal matrix. The two equations

$$\begin{bmatrix} 5 & -3 \\ -3 & 5 \end{bmatrix} \begin{bmatrix} 1/\sqrt{2} \\ 1/\sqrt{2} \end{bmatrix} = 2 \begin{bmatrix} 1/\sqrt{2} \\ 1/\sqrt{2} \end{bmatrix}$$

and

$$\begin{bmatrix} 5 & -3 \\ -3 & 5 \end{bmatrix} \begin{bmatrix} 1/\sqrt{2} \\ -1/\sqrt{2} \end{bmatrix} = 8 \begin{bmatrix} 1/\sqrt{2} \\ -1/\sqrt{2} \end{bmatrix}$$

can be combined to give the single equation

$$\begin{bmatrix} 5 & -3 \\ -3 & 5 \end{bmatrix} \begin{bmatrix} 1/\sqrt{2} & 1/\sqrt{2} \\ 1/\sqrt{2} & -1/\sqrt{2} \end{bmatrix} = \begin{bmatrix} 1/\sqrt{2} & 1/\sqrt{2} \\ 1/\sqrt{2} & -1/\sqrt{2} \end{bmatrix} \begin{bmatrix} 2 & 0 \\ 0 & 8 \end{bmatrix},$$

or $\boldsymbol{AP} = \boldsymbol{P}\Lambda$, where $\Lambda = \begin{bmatrix} 2 & 0 \\ 0 & 8 \end{bmatrix}$.

$$\boldsymbol{P}^{-1} = \boldsymbol{P}^{\mathrm{t}} \qquad \text{and} \qquad \boldsymbol{P}^{\mathrm{t}}\boldsymbol{AP} = \Lambda.$$

Now make the substitution $\boldsymbol{x} = \boldsymbol{Py}$, and the equation of the conic becomes

$$\boldsymbol{y}^{\mathrm{t}}\Lambda\boldsymbol{y} = 1,$$

or

$$2y_1^2 + 8y_2^2 = 1.$$

This is an ellipse with principal axes of length $2/\sqrt{2}$ and $2/\sqrt{8}$, i.e., $\sqrt{2}$ and $1/\sqrt{2}$. To find the equations of the axes we note that

$$\boldsymbol{y} = \boldsymbol{P}^{-1}\boldsymbol{x} = \boldsymbol{P}^{\mathrm{t}}\boldsymbol{x} = \begin{bmatrix} 1/\sqrt{2} & 1/\sqrt{2} \\ 1/\sqrt{2} & -1/\sqrt{2} \end{bmatrix} \boldsymbol{x}.$$

Thus

$$y_1 = (x_1 + x_2)/\sqrt{2},$$
$$y_2 = (x_1 - x_2)/\sqrt{2}.$$

The axes are given by $y_2 = 0$, $y_1 = 0$ respectively, or $x_1 - x_2 = 0$, $x_1 + x_2 = 0$. Thus the major axis ($y_2 = 0$) has equation $x_1 - x_2 = 0$ and the minor axis $x_1 + x_2 = 0$.

(b) The conic has equation $\boldsymbol{x}^{\mathrm{t}}\boldsymbol{Ax} = 4$, where $\boldsymbol{A} = \begin{bmatrix} 1 & 6 \\ 6 & 6 \end{bmatrix}$. The eigenvalues are given by

$$(1 - \lambda)(6 - \lambda) - 36 = 0,$$
$$\lambda^2 - 7\lambda - 30 = 0,$$
$$\lambda = 10 \text{ or } -3.$$

Eigenvectors corresponding to 10, -3 are $\{2, 3\}$, $\{3, -2\}$ respectively. Normalizing these by dividing each by $\sqrt{13}$ we obtain

$$\boldsymbol{P} = \begin{bmatrix} 2/\sqrt{13} & 3/\sqrt{13} \\ 3/\sqrt{13} & -2/\sqrt{13} \end{bmatrix}.$$

Writing $\boldsymbol{x} = \boldsymbol{P}\boldsymbol{y}$, the equation of the conic becomes

$$10y_1^2 - 3y_2^2 = 4.$$

This is a hyperbola with axes of lengths $4/\sqrt{10}$, $4/\sqrt{3}$. The equations of these axes are $3x_1 - 2x_2 = 0$, $2x_1 + 3x_2 = 0$.

These methods can readily be extended to a euclidean space of dimension 3, and the problem becomes that of determining the axes of a quadric.

An important problem in rigid dynamics is the determination of the principal axes of inertia of a given body at a given point O. If we take a set of rectangular cartesian axes Ox_1, Ox_2, Ox_3 we can calculate the moments of inertia A, B, C of the body about these axes and the products of inertia F, G, H with respect to the planes of reference. The ellipsoid

$$Ax_1^2 + Bx_2^2 + Cx_3^2 - 2Fx_2x_3 - 2Gx_3x_1 - 2Hx_1x_2 = 1,$$

called the momental ellipsoid at O, has the following property. If P is any point on it, the moment of inertia of the body about OP is inversely proportional to OP^2. If the equation of the ellipsoid is referred to its principal axes Oy_1, Oy_2, Oy_3 as axes of reference, the coefficients of y_2y_3, y_3y_1, y_1y_2 vanish and the products of inertia about the principal planes in pairs are all zero. These axes are called the principal axes of inertia of the body at O.

Example 2. The momental ellipsoid of a body at O, referred to rectangular cartesian axes Ox_1, Ox_2, Ox_3, has equation

$$5x_1^2 + 6x_2^2 + 7x_3^2 - 4x_1x_2 + 4x_2x_3 = 1.$$

Find the equations of its principal axes at O.

The equation of the ellipsoid may be written $\boldsymbol{x}^{\mathrm{t}}\boldsymbol{A}\boldsymbol{x} = 1$, where

$$\boldsymbol{A} = \begin{bmatrix} 5 & -2 & 0 \\ -2 & 6 & 2 \\ 0 & 2 & 7 \end{bmatrix}.$$

The eigenvalues are given by

$$\begin{bmatrix} 5-\lambda & -2 & 0 \\ -2 & 6-\lambda & 2 \\ 0 & 2 & 7-\lambda \end{bmatrix} = 0,$$

$$\lambda = 3,\ 6,\ \text{or } 9.$$

Corresponding eigenvectors are $\{2, 2, -1\}$, $\{2, -1, 2\}$, $\{-1, 2, 2\}$ respectively. We normalize each by dividing each by 3, and form the matrix $\boldsymbol{P}$ having these vectors as its columns. Thus

$$\boldsymbol{P} = \begin{bmatrix} 2/3 & 2/3 & -1/3 \\ 2/3 & -1/3 & 2/3 \\ -1/3 & 2/3 & 2/3 \end{bmatrix}$$

and

$$\boldsymbol{P}^{t}\boldsymbol{A}\boldsymbol{P} = \begin{bmatrix} 3 & 0 & 0 \\ 0 & 6 & 0 \\ 0 & 0 & 9 \end{bmatrix}.$$

Transform to new axes Oy_1, Oy_2, Oy_3 where $\boldsymbol{x} = \boldsymbol{P}\boldsymbol{y}$, and the equation of the ellipsoid becomes

$$3y_1^2 + 6y_2^2 + 9y_3^2 = 1.$$

Now $\boldsymbol{y} = \boldsymbol{P}^{-1}\boldsymbol{x} = \boldsymbol{P}^{t}\boldsymbol{x}$, so that

$$\begin{aligned} y_1 &= \tfrac{1}{3}(2x_1 + 2x_2 - x_3), \\ y_2 &= \tfrac{1}{3}(2x_1 - x_2 + 2x_3), \\ y_3 &= \tfrac{1}{3}(-x_1 + 2x_2 + 2x_3). \end{aligned}$$

In the new coordinate system the principal axes have equations $y_2 = y_3 = 0$, $y_3 = y_1 = 0$, $y_1 = y_2 = 0$. It follows that, in the original system, the principal axes of inertia are the intersections in pairs of the planes

$$\begin{aligned} 2x_1 + 2x_2 - x_3 &= 0, \\ 2x_1 - x_2 + 2x_3 &= 0, \\ -x_1 + 2x_2 + 2x_3 &= 0. \end{aligned}$$

Example 3. Find the nature of the quadric

$$5x_1^2 + 11x_2^2 - 2x_3^2 + 12x_1x_3 + 12x_2x_3 = 1$$

and the equations of its principal axes.

The quadric has equation $\boldsymbol{x}^{t}\boldsymbol{A}\boldsymbol{x} = 1$, where

$$\boldsymbol{A} = \begin{bmatrix} 5 & 0 & 6 \\ 0 & 11 & 6 \\ 6 & 6 & -2 \end{bmatrix}.$$

The eigenvalues are given by

$$\lambda^3 - 14\lambda^2 - 49\lambda + 686 = 0,$$
$$\lambda = 7,\ 14,\ -7.$$

The corresponding eigenvectors are $\{6, -3, 2\}$, $\{2, 6, 3\}$, $\{3, 2, -6\}$. In this case

$$\boldsymbol{P} = \begin{bmatrix} 6/7 & 2/7 & 3/7 \\ -3/7 & 6/7 & 2/7 \\ 2/7 & 3/7 & -6/7 \end{bmatrix},$$

and

$$\boldsymbol{P}^{t}\boldsymbol{A}\boldsymbol{P} = \begin{bmatrix} 7 & 0 & 0 \\ 0 & 14 & 0 \\ 0 & 0 & -7 \end{bmatrix}.$$

If we make the transformation $\boldsymbol{x} = \boldsymbol{P}\boldsymbol{y}$, the equation of the quadric becomes

$$7y_1^2 + 14y_2^2 - 7y_3^2 = 1.$$

This is a hyperboloid of one sheet. Now $\boldsymbol{y} = \boldsymbol{P}^{-1}\boldsymbol{x} = \boldsymbol{P}^{t}\boldsymbol{x}$, or

$$\begin{bmatrix} y_1 \\ y_2 \\ y_3 \end{bmatrix} = \begin{bmatrix} 6/7 & -3/7 & 2/7 \\ 2/7 & 6/7 & 3/7 \\ 3/7 & 2/7 & -6/7 \end{bmatrix} \begin{bmatrix} x_1 \\ x_2 \\ x_3 \end{bmatrix}.$$

The principal planes have equations

$$6x_1 - 3x_2 + 2x_3 = 0,$$
$$2x_1 + 6x_2 + 3x_3 = 0,$$
$$3x_1 + 2x_2 - 6x_3 = 0,$$

and the principal axes are the lines of intersection of these planes in pairs.

Here is an example to illustrate the procedure when the characteristic equation has a multiple root. In this case the corresponding eigenspace has dimension greater than 1, and we must find an orthonormal basis for this eigenspace.

Example 4. Reduce the quadratic form

$$2x_1^2 + 5x_2^2 + 2x_3^2 + 4x_2x_3 + 2x_3x_1 + 4x_1x_2$$

to a sum of squares by an orthogonal transformation.

The form can be written $\boldsymbol{x}^t A \boldsymbol{x}$, where

$$A = \begin{bmatrix} 2 & 2 & 1 \\ 2 & 5 & 2 \\ 1 & 2 & 2 \end{bmatrix}.$$

The characteristic equation is

$$(\lambda - 7)(\lambda - 1)^2 = 0.$$

Thus the eigenvalues are 7, 1, 1. The eigenvector corresponding to $\lambda = 7$ satisfies

$$-5x_1 + 2x_2 + x_3 = 0,$$
$$2x_1 - 2x_2 + 2x_3 = 0,$$

giving $\boldsymbol{x} = \{1, 2, 1\}$, or $\boldsymbol{x} = \{1/\sqrt{6}, 2/\sqrt{6}, 1/\sqrt{6}\}$ in normalized form. For $\lambda = 1$,

$$x_1 + 2x_2 + x_3 = 0$$

is the only equation obtained for the determination of the eigenvector. The solutions of this equation form a two-dimensional subspace and $\{2, -1, 0\}$, $\{1, 0, -1\}$ are clearly linearly independent and form a basis for this space. They are not, however, orthogonal. Take one of them, say $\{1, 0, -1\}$ as a basis vector, and take $\{1, 0, -1\} + k\{2, -1, 0\}$ as the second basis vector, choosing k so as to make this orthogonal to the first one. Thus

$$1 \,.\, (1 + 2k) + 0 \,.\, (0 - k) - 1 \,.\, (-1 + 0k) = 0,$$
$$k = -1.$$

This gives $\{1, 0, -1\}$ and $\{-1, 1, -1\}$ as an orthogonal basis. Finally, $\{1/\sqrt{2}, 0, -1/\sqrt{2}\}$ and $\{-1/\sqrt{3}, 1/\sqrt{3}, -1/\sqrt{3}\}$ form an orthonormal basis. Each of these is automatically orthogonal to the eigenvector corresponding to the different eigenvalue 7. Hence these two, together with $\{1/\sqrt{6}, 2/\sqrt{6}, 1/\sqrt{6}\}$ form an orthonormal basis for $\boldsymbol{R}_3$ consisting entirely of eigenvectors of $\boldsymbol{A}$. We now write $\boldsymbol{x} = \boldsymbol{P}\boldsymbol{y}$, where

$$\boldsymbol{P} = \begin{bmatrix} 1/\sqrt{6} & 1/\sqrt{2} & -1/\sqrt{3} \\ 2/\sqrt{6} & 0 & 1/\sqrt{3} \\ 1/\sqrt{6} & -1/\sqrt{2} & -1/\sqrt{3} \end{bmatrix},$$

and then $$\boldsymbol{P}^t\boldsymbol{A}\boldsymbol{P} = \begin{bmatrix} 7 & 0 & 0 \\ 0 & 1 & 0 \\ 0 & 0 & 1 \end{bmatrix}.$$

$\boldsymbol{P}$ is orthogonal, and under this transformation the quadratic form is reduced to

$$7y_1^2 + y_2^2 + y_3^2.$$

To conclude this chapter, here is an example of an application to dynamics, namely to the determination of the normal modes of oscillation of a vibrating system.

Example 5. An elastic string of natural length $3a$ and modulus μ has its ends attached to two points at a distance $6a$ apart on a smooth horizontal table. Equal masses m are attached to the string at its points of trisection and rest on the table. Find the normal modes of longitudinal vibration.

Suppose the masses are displaced along the line of the string and then released. Let x_1, x_2 respectively be their distances from their equilibrium position, measured in the same sense, at any instant. It is easy to verify that the equations of motion of the two masses are

$$m\ddot{x}_1 = \mu(-2x_1 + x_2)/a,$$
$$m\ddot{x}_2 = \mu(x_1 - 2x_2)/a.$$

If we write $\boldsymbol{x} = \{x_1, x_2\}$, $\ddot{\boldsymbol{x}} = \{\ddot{x}_1, \ddot{x}_2\}$, these two equations can be combined and written in matrix form

$$\ddot{\boldsymbol{x}} = \frac{\mu}{ma}\begin{bmatrix} -2 & 1 \\ 1 & -2 \end{bmatrix}\boldsymbol{x} = \frac{\mu}{ma}\boldsymbol{A}\boldsymbol{x},$$

where $\boldsymbol{A} = \begin{bmatrix} -2 & 1 \\ 1 & -2 \end{bmatrix}$ is a real symmetric matrix.

The eigenvalues of A are -1 and -3, and the corresponding eigenvectors $\{1/\sqrt{2}, 1/\sqrt{2}\}$ and $\{1/\sqrt{2}, -1/\sqrt{2}\}$. Let $P = \begin{bmatrix} 1/\sqrt{2} & 1/\sqrt{2} \\ 1/\sqrt{2} & -1/\sqrt{2} \end{bmatrix}$, so that $P^{t}P = I$. Then $P^{t}AP = \begin{bmatrix} -1 & 0 \\ 0 & -3 \end{bmatrix}$.

Now, let $x = Py$; the equation of motion then becomes

$$P\ddot{y} = \frac{\mu}{ma} APy.$$

Hence
$$P^{t}P\ddot{y} = \frac{\mu}{ma} P^{t}APy,$$

i.e.,
$$\ddot{y} = \frac{\mu}{ma}\begin{bmatrix} -1 & 0 \\ 0 & -3 \end{bmatrix} y.$$

Thus
$$\ddot{y}_1 = -\frac{\mu}{ma} y_1 \quad \text{and} \quad \ddot{y}_2 = -\frac{3\mu}{ma} y_2.$$

It follows that if $y_2 = 0$ throughout the motion, the oscillation is simple harmonic with period $2\pi\sqrt{\left(\frac{ma}{\mu}\right)}$, and if $y_1 = 0$ throughout, the oscillation is again simple harmonic with period $2\pi\sqrt{\left(\frac{ma}{3\mu}\right)}$. These are the fundamental periods and the two modes of vibration are the normal modes. But

$$y = P^{-1}x = P^{t}x = \begin{bmatrix} 1/\sqrt{2} & 1/\sqrt{2} \\ 1/\sqrt{2} & -1/\sqrt{2} \end{bmatrix} x,$$

so that
$$y_1 = (x_1 + x_2)/\sqrt{2},$$
$$y_2 = (x_1 - x_2)/\sqrt{2}.$$

Hence the normal modes are those in which $x_1 = x_2$ and the displacements of the two masses are the same throughout the motion, and $x_1 = -x_2$ and the displacements of the two masses are equal and opposite throughout the motion.

Problems

9.1 Determine which of the following forms is positive definite;

(a) $2x_1^2 + 3x_2^2 + 13x_3^2 + 10x_2x_3 - 4x_3x_1$,

(b) $3x_1^2 + 8x_2^2 - 24x_3^2 + 10x_2x_3 + 10x_3x_1 - 14x_1x_2$,

(c) $4x_1^2 + 10x_2^2 + 2x_3^2 - 8x_2x_3 - 4x_3x_1 + 4x_1x_2$.

Find non-zero values of the variables for which the form (c) vanishes.

9.2 Prove that $ax^2 + 2hxy + by^2$ is positive definite if and only if $a > 0$ and $ab - h^2 > 0$.

9.3 Find the rank and signature of each of the forms

(a) $9x_1^2 - 6x_1x_2 + 76x_2^2$,
(b) $4x_1^2 - 4x_1x_2 + x_2^2$,
(c) $2x_1^2 - 22x_2^2 + 9x_3^2 - 16x_2x_3 - 10x_3x_1 + 4x_1x_2$.

9.4 Find a matrix $\boldsymbol{P}$ such that $\boldsymbol{P}^t\boldsymbol{A}\boldsymbol{P}$ is diagonal if

$$\text{(a) } \boldsymbol{A} = \begin{bmatrix} 5 & -6 \\ -6 & 0 \end{bmatrix}, \qquad \text{(b) } \boldsymbol{A} = \begin{bmatrix} 0 & 2 & 0 \\ 2 & 0 & 1 \\ 0 & 1 & 0 \end{bmatrix}.$$

9.5 By means of an orthogonal transformation reduce to a sum of squares the quadratic forms;

(a) $x_1^2 + 2\sqrt{2}x_1x_2 + 2x_2^2$,
(b) $11x_1^2 + 5x_2^2 - 2x_3^2 + 12x_2x_3 + 12x_3x_1$,
(c) $-5x_1^2 + 7x_2^2 + 7x_3^2 + 32x_2x_3 - 16x_3x_1 - 16x_1x_2$.

9.6 By means of a unitary transformation reduce to a sum of squares the hermitian form

$$x_1\bar{x}_1 + (1 - i)x_1\bar{x}_2 + (1 + i)\bar{x}_1x_2 + 2x_2\bar{x}_2.$$

9.7 If A is a real symmetric $n \times n$ matrix, prove that there always exists an integer k such that $k\boldsymbol{I}_n + \boldsymbol{A}$ is positive definite. Find the smallest such k when $\boldsymbol{A} = \begin{bmatrix} 1 & 6 \\ 6 & 6 \end{bmatrix}$.

9.8 $\boldsymbol{A}$ is a real symmetric 2×2 matrix. Prove that the largest eigenvalue of $\boldsymbol{A}$ is equal to the largest value of $\boldsymbol{x}^t\boldsymbol{A}\boldsymbol{x}$ for all unit vectors $\boldsymbol{x}$ in the euclidean space $\boldsymbol{R}_2$.

10 · *Simultaneous Reduction of Two Forms*

We now investigate the possibility of reducing two hermitian forms to sums of squares by the same transformation. This cannot always be done, but it can under certain conditions which are often satisfied in practice. For example, if one of the forms is positive definite it can always be done. This has an important application in mechanics, for if one of the forms represents the kinetic energy of a system it is certainly positive definite.

10.1 Two methods of simultaneous reduction

We prove two theorems.

THEOREM 10.1. The two hermitian forms $\bar{\boldsymbol{x}}^{\mathrm{t}}\boldsymbol{A}\boldsymbol{x}$ and $\bar{\boldsymbol{x}}^{\mathrm{t}}\boldsymbol{B}\boldsymbol{x}$ in n variables $x_1, x_2, \ldots, x_n$ can be reduced simultaneously to sums of squares if one of the forms is positive definite.

Proof. Suppose $\bar{\boldsymbol{x}}^{\mathrm{t}}\boldsymbol{B}\boldsymbol{x}$ is positive definite. Then there exists a non-singular matrix $\boldsymbol{P}$ such that $\bar{\boldsymbol{P}}^{\mathrm{t}}\boldsymbol{B}\boldsymbol{P} = \boldsymbol{I}$ (see Section 9.3). Now if $\boldsymbol{A}$ is hermitian, $\boldsymbol{C} = \bar{\boldsymbol{P}}^{\mathrm{t}}\boldsymbol{A}\boldsymbol{P}$ is also hermitian. By Theorem 9.4, Corollary 1, there exists a unitary matrix $\boldsymbol{Q}$ such that $\bar{\boldsymbol{Q}}^{\mathrm{t}}\boldsymbol{C}\boldsymbol{Q} = \Lambda$, where Λ is diagonal. Λ is clearly hermitian and is therefore real. But $\bar{\boldsymbol{Q}}^{\mathrm{t}}\boldsymbol{I}\boldsymbol{Q} = \bar{\boldsymbol{Q}}^{\mathrm{t}}\boldsymbol{Q} = \boldsymbol{I}$, since $\boldsymbol{Q}$ is unitary. If we now apply the transformation

$\boldsymbol{x} = \boldsymbol{PQy}$, where $\boldsymbol{PQ}$ is non-singular since $\boldsymbol{P}$, $\boldsymbol{Q}$ are both non-singular, we obtain

$$\bar{\boldsymbol{x}}^{\mathrm{t}}\boldsymbol{A}\boldsymbol{x} = \bar{\boldsymbol{y}}^{\mathrm{t}}\bar{\boldsymbol{Q}}^{\mathrm{t}}\bar{\boldsymbol{P}}^{\mathrm{t}}\boldsymbol{APQy} = \bar{\boldsymbol{y}}^{\mathrm{t}}\bar{\boldsymbol{Q}}^{\mathrm{t}}\boldsymbol{CQy}$$

$$= \bar{\boldsymbol{y}}^{\mathrm{t}}\Lambda\boldsymbol{y} = \lambda_1|y_1|^2 + \lambda_2|y_2|^2 + \cdots + \lambda_n|y_n|^2$$

and

$$\bar{\boldsymbol{x}}^{\mathrm{t}}\boldsymbol{B}\boldsymbol{x} = \bar{\boldsymbol{y}}^{\mathrm{t}}\bar{\boldsymbol{Q}}^{\mathrm{t}}\bar{\boldsymbol{P}}^{\mathrm{t}}\boldsymbol{BPQy} = \bar{\boldsymbol{y}}^{\mathrm{t}}\bar{\boldsymbol{Q}}^{\mathrm{t}}\boldsymbol{IQy}$$

$$= \bar{\boldsymbol{y}}^{\mathrm{t}}\boldsymbol{y} = |y_1|^2 + |y_2|^2 + \cdots + |y_n|^2.$$

We have therefore reduced both forms to sums of squares.

Corollary. If $\boldsymbol{A}$, $\boldsymbol{B}$, and $\boldsymbol{x}$ are real, then $\boldsymbol{A}$ and $\boldsymbol{B}$ are symmetric and $\boldsymbol{x}^{\mathrm{t}}\boldsymbol{Ax}$, $\boldsymbol{x}^{\mathrm{t}}\boldsymbol{Bx}$ are quadratic forms. If $\boldsymbol{x}^{\mathrm{t}}\boldsymbol{Bx}$ is positive definite, then there exists a real non-singular matrix $\boldsymbol{P}$ such that $\boldsymbol{P}^{\mathrm{t}}\boldsymbol{BP} = \boldsymbol{I}$ (see Section 9.2). $\boldsymbol{C} = \boldsymbol{P}^{\mathrm{t}}\boldsymbol{AP}$ is symmetric, and there exists an orthogonal matrix $\boldsymbol{Q}$ such that $\boldsymbol{Q}^{\mathrm{t}}\boldsymbol{CQ} = \Lambda$. If we apply the non-singular transformation $\boldsymbol{x} = \boldsymbol{PQy}$, we obtain

$$\boldsymbol{x}^{\mathrm{t}}\boldsymbol{Ax} = \lambda_1 y_1^2 + \lambda_2 y_2^2 + \cdots + \lambda_n y_n^2,$$

$$\boldsymbol{x}^{\mathrm{t}}\boldsymbol{Bx} = y_1^2 + y_2^2 + \cdots + y_n^2.$$

THEOREM 10.2. The two hermitian forms $\bar{\boldsymbol{x}}^{\mathrm{t}}\boldsymbol{Ax}$ and $\bar{\boldsymbol{x}}^{\mathrm{t}}\boldsymbol{Bx}$ in n variables $x_1, x_2, \ldots, x_n$ can be reduced simultaneously to sums of squares if the equation $\det(\boldsymbol{A} - \lambda\boldsymbol{B}) = 0$ has n distinct roots.

Proof. Let the equation $\det(\boldsymbol{A} - \lambda\boldsymbol{B}) = 0$ have the n *distinct* roots $\lambda_1, \lambda_2, \ldots, \lambda_n$. We first show that these roots are real. For the matrix $\boldsymbol{A} - \lambda\boldsymbol{B}$ is singular $(1 \leqslant k \leqslant n)$ and hence, by Theorem 5.1, Corollary 2, there exists a vector $\boldsymbol{p}_k \neq \boldsymbol{0}$ such that

$$(\boldsymbol{A} - \lambda_k\boldsymbol{B})\boldsymbol{p}_k = \boldsymbol{0}.$$

Thus

$$\bar{\boldsymbol{p}}_k^{\mathrm{t}}\boldsymbol{A}\boldsymbol{p}_k = \lambda_k\bar{\boldsymbol{p}}_k^{\mathrm{t}}\boldsymbol{B}\boldsymbol{p}_k.$$

But each of the hermitian forms $\bar{\boldsymbol{p}}_k^{\mathrm{t}}\boldsymbol{A}\boldsymbol{p}_k$ and $\bar{\boldsymbol{p}}_k^{\mathrm{t}}\boldsymbol{B}\boldsymbol{p}_k$ is real, and so λ_k is real. Now

$$\boldsymbol{A}\boldsymbol{p}_k = \lambda_k\boldsymbol{B}\boldsymbol{p}_k \qquad (1 \leqslant k \leqslant n)$$

and taking the conjugate transpose of each side we obtain

$$\bar{\boldsymbol{p}}_k^{\mathrm{t}}\boldsymbol{A} = \lambda_k\bar{\boldsymbol{p}}_k^{\mathrm{t}}\boldsymbol{B} \qquad (1 \leqslant k \leqslant n),$$

since $\bar{\boldsymbol{A}}^{\mathrm{t}} = \boldsymbol{A}$, $\bar{\boldsymbol{B}}^{\mathrm{t}} = \boldsymbol{B}$, and λ_k is real. Hence

$$\bar{\boldsymbol{p}}_k^{\mathrm{t}} \boldsymbol{A} \boldsymbol{p}_h = \lambda_k \bar{\boldsymbol{p}}_k \boldsymbol{B} \boldsymbol{p}_h$$

and also $$\bar{\boldsymbol{p}}_k^{\mathrm{t}} \boldsymbol{A} \boldsymbol{p}_h = \lambda_h \bar{\boldsymbol{p}}_k^{\mathrm{t}} \boldsymbol{B} \boldsymbol{p}_h.$$

Thus $$(\lambda_k - \lambda_h) \bar{\boldsymbol{p}}_k^{\mathrm{t}} \boldsymbol{B} \boldsymbol{p}_h = 0.$$

But $$\lambda_k - \lambda_h \neq 0 \quad \text{if} \quad k \neq h.$$

Hence $$\bar{\boldsymbol{p}}_k^{\mathrm{t}} \boldsymbol{B} \boldsymbol{p}_h = 0 \quad \text{if} \quad k \neq h.$$

It follows that $$\bar{\boldsymbol{p}}_k^{\mathrm{t}} \boldsymbol{A} \boldsymbol{p}_h = 0 \quad \text{if} \quad k \neq h.$$

Write $\bar{\boldsymbol{p}}_k^{\mathrm{t}} \boldsymbol{A} \boldsymbol{p}_k = \alpha_k$, $\boldsymbol{p}_k^{\mathrm{t}} \boldsymbol{B} \boldsymbol{p}_k = \beta_k$ $(k = 1, \ldots, n)$. Note that the α_k and β_k are all real and $\alpha_k = \lambda_k \beta_k$.

Let $\boldsymbol{P}$ be the matrix having the vectors $\boldsymbol{p}_1, \boldsymbol{p}_2, \ldots, \boldsymbol{p}_n$ as its columns. The above equations can then be written in the form

$$\bar{\boldsymbol{P}}^{\mathrm{t}} \boldsymbol{A} \boldsymbol{P} = \begin{bmatrix} \alpha_1 & & & \\ & \alpha_2 & & \mathbf{O} \\ & & \cdot & \\ \mathbf{O} & & & \cdot \\ & & & & \alpha_n \end{bmatrix},$$

$$\bar{\boldsymbol{P}}^{\mathrm{t}} \boldsymbol{B} \boldsymbol{P} = \begin{bmatrix} \beta_1 & & & \\ & \beta_2 & & \mathbf{O} \\ & & \cdot & \\ \mathbf{O} & & & \cdot \\ & & & & \beta_n \end{bmatrix}.$$

Applying the transformation $\boldsymbol{x} = \boldsymbol{P}\boldsymbol{y}$ we obtain

$$\bar{\boldsymbol{x}}^{\mathrm{t}} \boldsymbol{A} \boldsymbol{x} = \bar{\boldsymbol{y}}^{\mathrm{t}} \bar{\boldsymbol{P}}^{\mathrm{t}} \boldsymbol{A} \boldsymbol{P} \boldsymbol{y} = \alpha_1 |y_1|^2 + \alpha_2 |y_2|^2 + \cdots + \alpha_n |y_n|^2,$$

$$\bar{\boldsymbol{x}}^{\mathrm{t}} \boldsymbol{B} \boldsymbol{x} = \bar{\boldsymbol{y}}^{\mathrm{t}} \bar{\boldsymbol{P}}^{\mathrm{t}} \boldsymbol{B} \boldsymbol{P} \boldsymbol{y} = \beta_1 |y_1|^2 + \beta_2 |y_2|^2 + \cdots + \beta_n |y_n|^2.$$

We can choose the vectors $\boldsymbol{p}_k$ in such a way as to make $\alpha_k = 0$, 1 or -1 for each k. For, if $\alpha_k > 0$ and $\boldsymbol{q}_k = \boldsymbol{p}_k / \sqrt{\alpha_k}$, $\bar{\boldsymbol{q}}_k^{\mathrm{t}} \boldsymbol{A} \boldsymbol{q}_k = 1$, and if $\alpha_k < 0$ and $\boldsymbol{q}_k = \boldsymbol{p}_k / \sqrt{(-\alpha_k)}$, $\bar{\boldsymbol{q}}_k^{\mathrm{t}} \boldsymbol{A} \boldsymbol{q}_k = -1$.

Corollary 1. The theorem holds in certain cases even when the roots of the equation $\det(\boldsymbol{A} - \lambda \boldsymbol{B}) = 0$ are not distinct.

Suppose that the root λ of the equation has multiplicity k, and suppose that the null space of the matrix $\boldsymbol{A} - \lambda \boldsymbol{B}$ (i.e., the kernel of the transformation with matrix $A - \lambda B$) has dimension k. This means that we can find k linearly independent vectors $q_1, q_2, \ldots, q_k$ such that

$$(A - \lambda B)q_h = 0 \qquad (1 \leqslant h \leqslant k).$$

Now, for each h $(1 \leqslant h \leqslant k)$, replace q_h by p_h, where p_h is a linear combination of the vectors $q_1, q_2, \ldots, q_h$. We can choose p_h so that

$$\bar{p}_j^t(A - \lambda B)p_h = 0 \qquad (j \neq h).$$

The process now continues as before and we shall illustrate this later by means of an example.

This result is analogous to that of Theorem 6.3. The method applies when the vectors p satisfying the equation $(A - \lambda B)p = 0$, where $\det(A - \lambda B) = 0$, span the space C_n.

Corollary 2. Let x^tAx and x^tBx be quadratic forms. As in the corollary to Theorem 10.1 we can show that, if the equation $\det(A - \lambda B) = 0$ has n distinct roots, then the two forms can be reduced simultaneously to sums of squares.

10.2 Examples and applications

In the following examples we are asked to perform the reduction for the two given forms.

Example 1. x^tAx and x^tBx, where

$$A = \begin{bmatrix} 5 & 2 \\ 2 & -2 \end{bmatrix}, \qquad B = \begin{bmatrix} -2 & 1 \\ 1 & -1 \end{bmatrix}.$$

Neither of these forms is positive definite. However,

$$\det(A - \lambda B) = \begin{vmatrix} 5 + 2\lambda & 2 - \lambda \\ 2 - \lambda & -2 + \lambda \end{vmatrix} = (\lambda + 7)(\lambda - 2).$$

Hence we can apply the method of Theorem 10.2.

$\lambda = -7$ gives the equation $\begin{bmatrix} -9 & 9 \\ 9 & -9 \end{bmatrix} p = 0,\ p = \{1, 1\}.$

$\lambda = 2$ gives the equation $\begin{bmatrix} 9 & 0 \\ 0 & 0 \end{bmatrix} p = 0,\ p = \{0, 1\}.$

Thus $P = \begin{bmatrix} 1 & 0 \\ 1 & 1 \end{bmatrix}, \quad P^tAP = \begin{bmatrix} 7 & 0 \\ 0 & -2 \end{bmatrix}, \quad P^tBP = \begin{bmatrix} -1 & 0 \\ 0 & -1 \end{bmatrix}.$

Writing $\boldsymbol{x} = \boldsymbol{P}\boldsymbol{y}$, we obtain

$$\boldsymbol{x}^t\boldsymbol{A}\boldsymbol{x} = \boldsymbol{y}^t\begin{bmatrix} 7 & 0 \\ 0 & -2 \end{bmatrix}\boldsymbol{y} = 7y_1^2 - 2y_2^2,$$

$$\boldsymbol{x}^t\boldsymbol{B}\boldsymbol{x} = \boldsymbol{y}^t\begin{bmatrix} -1 & 0 \\ 0 & -1 \end{bmatrix}\boldsymbol{y} = -y_1^2 - y_2^2.$$

The required transformation is $\boldsymbol{x} = \boldsymbol{P}\boldsymbol{y}$, or

$$x_1 = y_1,$$

$$x_2 = y_1 + y_2.$$

Example 2. $\bar{\boldsymbol{x}}^t\boldsymbol{A}\boldsymbol{x}$ and $\bar{\boldsymbol{x}}^t\boldsymbol{B}\boldsymbol{x}$, where

$$\boldsymbol{A} = \begin{bmatrix} 1 & 1+i \\ 1-i & 2 \end{bmatrix}, \qquad \boldsymbol{B} = \begin{bmatrix} -3 & 2-5i \\ 2+5i & 1 \end{bmatrix}.$$

$$\det(\boldsymbol{A} - \lambda\boldsymbol{B}) = \begin{vmatrix} 1+3\lambda & 1-2\lambda+i(1+5\lambda) \\ 1-2\lambda-i(1+5\lambda) & 2-\lambda \end{vmatrix}$$

$$= 32\lambda^2 + \lambda.$$

$\lambda = 0$ gives $\boldsymbol{A} - \lambda\boldsymbol{B} = \boldsymbol{A}$, and we solve the equation $\boldsymbol{A}\boldsymbol{p}_1 = \boldsymbol{0}$. $\boldsymbol{p}_1 = \{1 + i, -1\}$.

$\lambda = -1/32$ gives $\boldsymbol{A} + \frac{1}{32}\boldsymbol{B}$, and we solve the equation $\left(\boldsymbol{A} + \frac{1}{32}\boldsymbol{B}\right)\boldsymbol{p}_2 = \boldsymbol{0}$, $\boldsymbol{p}_2 = \{34 + 27i, -29\}$.

We now make the transformation $\boldsymbol{x} = \boldsymbol{P}\boldsymbol{y}$, where

$$\boldsymbol{P} = \begin{bmatrix} 1+i & 34+27i \\ -1 & -29 \end{bmatrix}.$$

Then $\quad \bar{\boldsymbol{P}}^t\boldsymbol{A}\boldsymbol{P} = \begin{bmatrix} 0 & 0 \\ 0 & 29 \end{bmatrix} \quad$ and $\quad \bar{\boldsymbol{P}}^t\boldsymbol{B}\boldsymbol{P} = \begin{bmatrix} 1 & 0 \\ 0 & -928 \end{bmatrix}.$

Thus $\quad \bar{\boldsymbol{x}}^t\boldsymbol{A}\boldsymbol{x} = \bar{\boldsymbol{y}}^t\bar{\boldsymbol{P}}^t\boldsymbol{A}\boldsymbol{P}\boldsymbol{y} = 29y_2^2,$

and $\quad \bar{\boldsymbol{x}}^t\boldsymbol{B}\boldsymbol{x} = \bar{\boldsymbol{y}}^t\bar{\boldsymbol{P}}^t\boldsymbol{B}\boldsymbol{P}\boldsymbol{y} = y_1^2 - 928y_2^2.$

Both hermitian forms are therefore reduced to sums of squares by the transformation

$$x_1 = (1 + i)y_1 + (34 + 27i)y_2,$$

$$x_2 = -y_1 - 29y_2.$$

Note that in this case also neither form is positive definite.

Example 3. $\boldsymbol{x}^t\boldsymbol{A}\boldsymbol{x}$ and $\boldsymbol{x}^t\boldsymbol{B}\boldsymbol{x}$, where

$$\boldsymbol{A} = \begin{bmatrix} 1 & -3 & 2 \\ -3 & 3 & 0 \\ 2 & 0 & 4 \end{bmatrix}, \qquad \boldsymbol{B} = \begin{bmatrix} 2 & 0 & 1 \\ 0 & 3 & 0 \\ 1 & 0 & 2 \end{bmatrix}.$$

$$\begin{aligned} \boldsymbol{x}^t\boldsymbol{B}\boldsymbol{x} &= 2x_1^2 + 3x_2^2 + 2x_3^2 + 2x_1x_3 \\ &= 2(x_1 + \tfrac{1}{2}x_3)^2 + 3x_2^2 + \tfrac{3}{2}x_3^2. \end{aligned}$$

Thus $\boldsymbol{x}^t\boldsymbol{B}\boldsymbol{x}$ is positive definite. Take

$$\begin{aligned} y_1 &= \sqrt{2}x_1 + x_3/\sqrt{2}, \\ y_2 &= \sqrt{3}x_2, \\ y_3 &= \frac{\sqrt{3}}{\sqrt{2}}x_3. \end{aligned}$$

Then

$$\begin{aligned} x_1 &= \frac{1}{\sqrt{2}}y_1 - \frac{1}{\sqrt{6}}y_3, \\ x_2 &= \frac{1}{\sqrt{3}}y_2, \\ x_3 &= \frac{\sqrt{2}}{\sqrt{3}}y_3, \end{aligned}$$

or $\boldsymbol{x} = \boldsymbol{P}\boldsymbol{y}$, where

$$\boldsymbol{P} = \begin{bmatrix} 1/\sqrt{2} & 0 & -1/\sqrt{6} \\ 0 & 1/\sqrt{3} & 0 \\ 0 & 0 & \sqrt{2}/\sqrt{3} \end{bmatrix} = \frac{1}{\sqrt{6}}\begin{bmatrix} \sqrt{3} & 0 & -1 \\ 0 & \sqrt{2} & 0 \\ 0 & 0 & 2 \end{bmatrix}$$

and $\boldsymbol{P}^t\boldsymbol{B}\boldsymbol{P} = \boldsymbol{I}$.

$$\boldsymbol{C} = \boldsymbol{P}^t\boldsymbol{A}\boldsymbol{P} = \begin{bmatrix} 1/2 & -\sqrt{3}/\sqrt{2} & \sqrt{3}/2 \\ -\sqrt{3}/\sqrt{2} & 1 & 1/\sqrt{2} \\ \sqrt{3}/2 & 1/\sqrt{2} & 3/2 \end{bmatrix}$$

$$= \frac{1}{2}\begin{bmatrix} 1 & -\sqrt{6} & \sqrt{3} \\ -\sqrt{6} & 2 & \sqrt{2} \\ \sqrt{3} & \sqrt{2} & 3 \end{bmatrix}.$$

The eigenvalues of C are given by

$$\begin{vmatrix} 1-2\lambda & -\sqrt{6} & \sqrt{3} \\ -\sqrt{6} & 2-2\lambda & \sqrt{2} \\ \sqrt{3} & \sqrt{2} & 3-2\lambda \end{vmatrix} = 0,$$

i.e.,

$$(\lambda+1)(\lambda-2)^2 = 0.$$

$\lambda = -1$ gives the eigenvector $\{\sqrt{3}, \sqrt{2}, -1\}$. $\lambda = 2$ gives the eigenvectors $\{0, 1, \sqrt{2}\}$, $\{1, 0, \sqrt{3}\}$. However, these two vectors are not orthogonal. Replace the second one by $\{1, 0, \sqrt{3}\} + k\{0, 1, \sqrt{2}\}$ and choose k to make this orthogonal to $\{0, 1, \sqrt{2}\}$. Thus

$$k + (\sqrt{3} + \sqrt{2}k)\sqrt{2} = 0,$$
$$k = -\sqrt{2}/\sqrt{3}.$$

This gives the vector $\{1, -\sqrt{2}/\sqrt{3}, \sqrt{3} - 2/\sqrt{3}\} = \dfrac{1}{\sqrt{3}}\{\sqrt{3}, -\sqrt{2}, 1\}$.

We now normalize the three vectors, giving the columns $\dfrac{1}{\sqrt{6}}\{\sqrt{3}, \sqrt{2}, -1\}$, $\dfrac{1}{\sqrt{3}}\{0, 1, \sqrt{2}\}$, $\dfrac{1}{\sqrt{6}}\{\sqrt{3}, -\sqrt{2}, 1\}$ of the matrix Q. Thus

$$Q = \begin{bmatrix} 1/\sqrt{2} & 0 & 1/\sqrt{2} \\ 1/\sqrt{3} & 1/\sqrt{3} & -1/\sqrt{3} \\ -1/\sqrt{6} & \sqrt{2}/\sqrt{3} & 1/\sqrt{6} \end{bmatrix}$$

and

$$Q^{t}CQ = \begin{bmatrix} -1 & 0 & 0 \\ 0 & 2 & 0 \\ 0 & 0 & 2 \end{bmatrix}.$$

Moreover

$$PQ = \tfrac{1}{3}\begin{bmatrix} 2 & -1 & 1 \\ 1 & 1 & -1 \\ -1 & 2 & 1 \end{bmatrix}.$$

Hence

$$x_1 = \tfrac{1}{3}(2y_1 - y_2 + y_3),$$
$$x_2 = \tfrac{1}{3}(y_1 + y_2 - y_3),$$
$$x_3 = \tfrac{1}{3}(-y_1 + 2y_2 + y_3),$$

transforms $\boldsymbol{x}^{t}A\boldsymbol{x}$ and $\boldsymbol{x}^{t}B\boldsymbol{x}$ into $-y_1^2 + 2y_2^2 + 2y_3^2$ and $y_1^2 + y_2^2 + y_3^2$ respectively.

Let us now do the same example by the second method. In this case the equation $\det(A - \lambda B) = 0$ is

$$(2-\lambda)^2(\lambda+1) = 0,$$

so that $\lambda = -1, 2, 2$. The roots are not distinct, but we can nevertheless continue by the method described in Theorem 10.2, Corollary 1. If $\lambda = -1$ and $(\boldsymbol{A} + \boldsymbol{B})\boldsymbol{p}_1 = \boldsymbol{0}$, we find on solving the equations that $\boldsymbol{p}_1 = \{2, 1, -1\}$. If $\lambda = 2$ and $(\boldsymbol{A} - 2\boldsymbol{B})\boldsymbol{p}_2 = \boldsymbol{0}$, then we can find two linearly independent solutions for $\boldsymbol{p}_2$; e.g., $\boldsymbol{p}_2 = \{1, -1, 0\}$ or $\boldsymbol{q}_2 = \{1, -1, 1\}$. Now $\boldsymbol{p}_1^t\boldsymbol{B}\boldsymbol{p}_2 = 0$ since $\boldsymbol{p}_1$, $\boldsymbol{p}_2$ belong to *different* values of λ. Similarly $\boldsymbol{p}_1^t\boldsymbol{B}\boldsymbol{q}_2 = 0$. But $\boldsymbol{p}_2^t\boldsymbol{B}\boldsymbol{q}_2 = 6 \neq 0$. Now $\boldsymbol{p}_2^t\,\boldsymbol{B}\boldsymbol{p}_2 = 5$, and hence

$\boldsymbol{p}_2^t\boldsymbol{B}\boldsymbol{p}_3 = 0$, where $\boldsymbol{p}_3 = 5\boldsymbol{q}_2 - 6\boldsymbol{p}_2 = \{-1, 1, 5\}$. $\boldsymbol{p}_1^t\boldsymbol{B}\boldsymbol{p}_1 = 9$ and $\boldsymbol{p}_3^t\boldsymbol{B}\boldsymbol{p}_3 = 45$.

The numbers $\alpha_1, \alpha_2, \alpha_3$ of the theorem are 9, 5, 45 respectively, and if we replace $\boldsymbol{p}_1, \boldsymbol{p}_2, \boldsymbol{p}_3$ by $\frac{1}{3}\boldsymbol{p}_1, \frac{1}{\sqrt{5}}\boldsymbol{p}_2, \frac{1}{3\sqrt{5}}\boldsymbol{p}_3$ we obtain the columns of a matrix $\boldsymbol{P}$ such that $\boldsymbol{P}^t\boldsymbol{B}\boldsymbol{P} = \boldsymbol{I}$. Now

$$\boldsymbol{P} = \begin{bmatrix} 2/3 & 1/\sqrt{5} & -1/3\sqrt{5} \\ 1/3 & -1/\sqrt{5} & 1/3\sqrt{5} \\ -1/3 & 0 & \sqrt{5}/3 \end{bmatrix},$$

so that the transformation $\boldsymbol{x} = \boldsymbol{P}\boldsymbol{y}$ is given by

$$x_1 = \frac{2}{3}y_1 + \frac{1}{\sqrt{5}}y_2 - \frac{1}{3\sqrt{5}}y_3,$$

$$x_2 = \frac{1}{3}y_1 - \frac{1}{\sqrt{5}}y_2 + \frac{1}{3\sqrt{5}}y_3,$$

$$x_3 = -\frac{1}{3}y_1 + \frac{\sqrt{5}}{3}y_3.$$

Under this transformation $\boldsymbol{x}^t\boldsymbol{B}\boldsymbol{x}$ becomes $y_1^2 + y_2^2 + y_3^2$ and $\boldsymbol{x}^t\boldsymbol{A}\boldsymbol{x}$ becomes $-y_1^2 + 2y_2^2 + 2y_3^2$. The transformation is not the same as the one we obtained last time.

The following example illustrates an application in mechanics.

Example 4. A smooth circular wire of radius a and mass $2m$ is free to rotate in a vertical plane about a fixed point O on its circumference. A bead of mass $3m$ is threaded on the wire and is free to slide on it. The system performs small oscillations. Find the normal modes and the principal periods.

At time t let the radius of the wire through O make an angle θ with the vertical and let the chord of the circle joining O to the bead make an angle ϕ with the vertical, both angles being measured in the same sense. θ and ϕ remain small throughout the motion. Clearly, the kinetic energy T and the potential energy V of the system are given by

$$T = \tfrac{1}{2} \,.\, 2m \,.\, 2a\dot{\theta}^2 + \tfrac{1}{2} \,.\, 3m\left\{\frac{d}{dt}[2a\cos(\theta - \phi)\cos\phi]\right\}^2$$

$$+ \tfrac{1}{2} \,.\, 3m\left\{\frac{d}{dt}[2a\cos(\theta - \phi)\sin\phi]\right\}^2,$$

$$V = 2mga(1 - \cos\theta) + 3mg \,.\, 2a[1 - \cos(\theta - \phi)\cos\phi]$$

where V is chosen in such a way that $V = 0$ in the equilibrium position. $L = T - V$ is the Lagrangian function, and Lagrange's equations of motion are then

$$\frac{d}{dt}\left(\frac{\partial L}{\partial \dot{\theta}}\right) - \frac{\partial L}{\partial \theta} = 0,$$

$$\frac{d}{dt}\left(\frac{\partial L}{\partial \dot{\phi}}\right) - \frac{\partial L}{\partial \phi} = 0.$$

In these equations we neglect terms of degree higher than the first in the variables θ, ϕ. The reader should satisfy himself that this is equivalent to neglecting terms of degree higher than the first in the expression for T, and neglecting terms of degree higher than the second in the expression for V. Replacing the sines and cosines by the appropriate terms of their series expansions we then obtain

$$T = 2ma^2\dot{\theta}^2 + 6ma^2\dot{\phi}^2,$$

$$V = mga\theta^2 + 6mga\left(\frac{\theta^2}{2} - \theta\phi + \phi^2\right)$$

$$= 2mga(2\theta^2 - 3\theta\phi + 3\phi^2).$$

If we denote by $\boldsymbol{x}$ the column vector $\{\theta, \phi\}$, T and V can be expressed as quadratic forms

$$T = 2ma^2 \,.\, \dot{\boldsymbol{x}}^t\boldsymbol{B}\dot{\boldsymbol{x}},$$

$$V = 2mga \,.\, \boldsymbol{x}^t\boldsymbol{A}\boldsymbol{x}$$

where $$\boldsymbol{B} = \begin{bmatrix} 1 & 0 \\ 0 & 3 \end{bmatrix}, \qquad \boldsymbol{A} = \begin{bmatrix} 2 & -3/2 \\ -3/2 & 3 \end{bmatrix}.$$

We now find a transformation of the variables that will transform these quadratic forms simultaneously to sums of squares.

$$\det(\boldsymbol{A} - \lambda\boldsymbol{B}) = 0 \qquad \text{gives} \qquad \begin{vmatrix} 2 - \lambda & -3/2 \\ -3/2 & 3 - 3\lambda \end{vmatrix} = 0,$$

$$\lambda = 1/2 \text{ or } 5/2.$$

With the notation used earlier,

$$\lambda = 1/2 \quad \text{gives} \quad \boldsymbol{p}_1 = \{1, 1\},$$
$$\lambda = 5/2 \quad \text{gives} \quad \boldsymbol{p}_2 = \{3, -1\},$$

$$\boldsymbol{P} = \begin{bmatrix} 1 & 3 \\ 1 & -1 \end{bmatrix},$$

$$\boldsymbol{P}^t\boldsymbol{A}\boldsymbol{P} = \begin{bmatrix} 2 & 0 \\ 0 & 30 \end{bmatrix}, \qquad \boldsymbol{P}^t\boldsymbol{B}\boldsymbol{P} = \begin{bmatrix} 4 & 0 \\ 0 & 12 \end{bmatrix}.$$

Let $\boldsymbol{x} = \boldsymbol{P}\boldsymbol{y}$, where $\boldsymbol{y} = \{\alpha, \beta\}$. Then

$$\theta = \alpha + 3\beta,$$
$$\phi = \alpha - \beta,$$

and
$$T = 2ma^2[4\dot{\alpha}^2 + 12\dot{\beta}^2] = 8ma^2[\dot{\alpha}^2 + 3\dot{\beta}^2],$$
$$V = 2mga[2\alpha^2 + 30\beta^2] = 4mga[\alpha^2 + 15\beta^2].$$

Lagrange's equations become

$$\frac{d}{dt}[16ma^2\dot{\alpha}] + 8mga\alpha = 0,$$

$$\frac{d}{dt}[48ma^2\dot{\beta}] + 120mga\beta = 0.$$

i.e.,
$$2a\ddot{\alpha} = -g\alpha,$$
$$2a\ddot{\beta} = -5g\beta.$$

Thus, the principal periods are $2\pi\sqrt{\left(\frac{2a}{g}\right)}$, $2\pi\sqrt{\left(\frac{2a}{5g}\right)}$. The normal modes are given by $\beta \equiv 0$, $\alpha \equiv 0$ respectively, i.e., $\theta - \phi \equiv 0$, $\theta + 3\phi \equiv 0$. In the first case the system oscillates as a pendulum, with the bead remaining fixed and diametrically opposite to O. In the second case the radius through the bead meets the vertical through O in a fixed point.

Problems

Transform the following quadratic forms simultaneously to sums of squares;

10.1 $5x_1^2 - 26x_1x_2 + 34x_2^2$,
$5x_1^2 - 22x_1x_2 + 23x_2^2$.

10.2 $4x_1^2 + 3x_2^2 + 5x_3^2 + 6x_2x_3 - 12x_3x_1 - 4x_1x_2$,
$8x_1^2 + 5x_2^2 + 6x_3^2 + 10x_2x_3 + 4x_1x_2$.

10.3 $9x_1^2 + 6x_2^2 + 8x_3^2 + 4x_2x_3 + 4x_3x_1 - 4x_1x_2$,
$5x_1^2 + 5x_2^2 + 12x_2x_3 - 12x_3x_1 + 8x_1x_2$.

10.4 $x_2^2 + x_3^2 + 2x_2x_3 + 4x_1x_2$,
$2x_1^2 + 3x_2^2 + 3x_3^2 - 2x_2x_3 + 4x_3x_1$.

10.5 Show that the form

$$f = 2x_1^2 + 3x_2^2 + 2x_3^2 + 2x_3x_1$$

is positive definite. If

$$g = x_1^2 + 3x_2^2 + 4x_3^2 + 4x_3x_1 - 6x_1x_2,$$

carry out the simultaneous reduction of f, g to $y_1^2 + y_2^2 + y_3^2$, $\lambda_1 y_1^2 + \lambda_2 y_2^2 + \lambda_3 y_3^2$ respectively.

10.6 An inextensible string of length $4a/5$ is attached at one end to a fixed point O and at the other to an end of a uniform rod of length $2a$. The system makes small oscillations in a vertical plane through O. Find the principal periods and show that in one of the normal modes the rod always passes through a point at a distance $2a$ vertically below O.

Suggested Reading

1. BIRKHOFF, G. and S. MACLANE. *A survey of modern algebra.* Macmillan, New York, 1950.
2. CULLEN, C. *Matrices and linear transformations.* Addison-Wesley, Reading, Mass., 1966.
3. CURTIS, C. W. *Linear algebra.* Allyn and Bacon, Boston, 1963.
4. FINKBEINER, D. T. *Introduction to matrices and linear transformations.* W. H. Freeman, San Francisco, 1960.
5. GRUENBERG, K. W. and A. J. WEIR. *Linear geometry.* Van Nostrand, Princeton, New Jersey, 1967.
6. HADLEY, G. *Linear algebra.* Addison-Wesley, Reading, Mass., 1965.
7. HALMOS, P. R. *Finite-dimensional vector spaces.* Van Nostrand, Princeton, New Jersey, 1958.
8. MIRSKY, L. *An introduction to linear algebra.* Clarendon Press, Oxford, 1955.
9. MOSTOW, G. D., J. H. SAMPSON and J. P. MEYER. *Fundamental structures of algebra.* McGraw-Hill, New York, 1963.
10. SCHREIER, O. and E. SPERNER. *Introduction to modern algebra and matrix theory.* Chelsea, New York, 1959.
11. SHEPHARD, G. C. *Vector spaces of finite dimension.* Oliver and Boyd, Edinburgh, 1965.
12. TROPPER, A. MARY. *Matrix theory for electrical engineering students.* Harrap, London, and Addison-Wesley, Reading, Mass., 1962.

Answers to problems

1.7 $0 + 0\sqrt{2}$, $1 + 0\sqrt{2}$.

1.8 $0 + 0\omega$, $1 + 0\omega$.

1.9 0, 1.

1.10 $0 = 0/1$, $1 = 1/1$.

1.11 No. The set is not closed under addition. E.g.,

$$\begin{bmatrix} 1 & 1 \\ 1 & 2 \end{bmatrix} + \begin{bmatrix} 1 & 2 \\ 1 & 1 \end{bmatrix} = \begin{bmatrix} 2 & 3 \\ 2 & 3 \end{bmatrix},$$

which is singular.

2.2 E.g., $S = [1, \sqrt{2}]$, $T = [\sqrt{5}]$.

2.3 $\{\boldsymbol{v}_1, \boldsymbol{v}_2, \boldsymbol{e}_1, \boldsymbol{e}_3\}$ is a basis.

2.4 $$\left\{\begin{bmatrix} 1 & 0 \\ 0 & 0 \end{bmatrix}, \begin{bmatrix} 0 & 1 \\ 0 & 0 \end{bmatrix}, \begin{bmatrix} 0 & 0 \\ 1 & 0 \end{bmatrix}, \begin{bmatrix} 0 & 0 \\ 0 & 1 \end{bmatrix}, \begin{bmatrix} i & 0 \\ 0 & 0 \end{bmatrix}, \begin{bmatrix} 0 & i \\ 0 & 0 \end{bmatrix}, \begin{bmatrix} 0 & 0 \\ i & 0 \end{bmatrix}, \begin{bmatrix} 0 & 0 \\ 0 & i \end{bmatrix}\right\}.$$

Dimension 8.

2.5 (1, 2, 1), (0, 1, −2).

2.6 One.

2.7 $k = -2$ or 3.

2.8 $\dim M = 3$, $\dim N = 3$,
$\dim (M \cap N) = 2$,
$\dim (M + N) = 4$.

2.11 $\{1, X, X^2, \ldots, X^n\}$.
Dimension $= n + 1$.

2.12 $$U = \left[\begin{bmatrix} 1 & 0 \\ 0 & 0 \end{bmatrix}, \begin{bmatrix} 0 & 0 \\ 0 & 1 \end{bmatrix}\right],$$
$$W = \left[\begin{bmatrix} 1 & 0 \\ 0 & 0 \end{bmatrix}, \begin{bmatrix} 0 & 0 \\ 1 & 0 \end{bmatrix}\right].$$
$$U \cap W = \left[\begin{bmatrix} 1 & 0 \\ 0 & 0 \end{bmatrix}\right],$$
$$U + W = \left[\begin{bmatrix} 1 & 0 \\ 0 & 0 \end{bmatrix}, \begin{bmatrix} 0 & 0 \\ 1 & 0 \end{bmatrix}, \begin{bmatrix} 0 & 0 \\ 0 & 1 \end{bmatrix}\right].$$
$\dim U = 2$, $\dim W = 2$,
$\dim (U \cap W) = 1$,
$\dim (U + W) = 3$.

2.13 $\dim U = 2$, $\dim V = 3$,
$\dim (U + V) = 4$,
$\dim (U \cap V) = 1$.

3.1 $\begin{bmatrix} 1 & 0 \\ 0 & -1 \end{bmatrix}$. $T^{-1} = T$.

3.2 $(n - p) \times p$.

3.3 $$\begin{bmatrix} -11 & -2 & 14 \\ 9 & 2 & -11 \\ -6 & 0 & 9 \end{bmatrix}.$$

3.4 $$\begin{bmatrix} -11/5 & 9/10 \\ 4/5 & 9/10 \end{bmatrix}.$$

3.5 $ST - TS$ is the identity mapping on P.

3.6 (i), (iii), (iv), (v) are linear.

3.7 $$\begin{aligned} x_1 &= 23y_1 + 10y_2 - 16y_3, \\ x_2 &= -7y_1 - 3y_2 + 5y_3, \\ x_3 &= 4y_1 + 2y_2 - 3y_3. \end{aligned}$$
$$A^{-1} = \begin{bmatrix} 23 & 10 & -16 \\ -7 & -3 & 5 \\ 4 & 2 & -3 \end{bmatrix}.$$

3.10 (*a*) $(-10, -3)$,
(*b*) $(6, -3)$,
(*c*) $(x_1 + 2x_2 - x_3, 2x_2 + 2x_3)$,
(*d*) $(0, 5)$.
Matrix of A is $\begin{bmatrix} 1 & 1 & -1 \\ 0 & 2 & 1 \end{bmatrix}$.
Matrix of B is $\begin{bmatrix} 0 & 1 & 0 \\ 0 & 0 & 1 \end{bmatrix}$.

3.11 $AB = BA = -B$.

4.1 $x = 7$, rank $= 3$.
$x \neq 7$, rank $= 4$.

4.4 (i) E.g.,

$$\boldsymbol{H} = \begin{bmatrix} 1 & 1 & 2 & 0 \\ -1 & -3 & 8 & 0 \\ 4 & -3 & -7 & 0 \\ 1 & 12 & -3 & 1 \end{bmatrix},$$

$$\boldsymbol{K} = \boldsymbol{I}_3,\ \boldsymbol{H}^{-1}\boldsymbol{A}\boldsymbol{K} = \begin{bmatrix} \boldsymbol{I}_3 \\ \boldsymbol{O} \end{bmatrix}.$$

(ii) E.g.,

$$\boldsymbol{K} = \begin{bmatrix} 1 & 0 & -5 & 6 \\ 0 & 1 & 1 & -5 \\ 0 & 0 & 1 & 0 \\ 0 & 0 & 0 & 1 \end{bmatrix},$$

$$\boldsymbol{H} = \begin{bmatrix} 1 & 2 & 0 \\ 0 & 1 & 0 \\ 3 & 4 & 1 \end{bmatrix},$$

$$\boldsymbol{H}^{-1}\boldsymbol{A}\boldsymbol{K} = \begin{bmatrix} \boldsymbol{I}_2 & \boldsymbol{0} \\ \boldsymbol{0} & \boldsymbol{0} \end{bmatrix}.$$

4.5 (i) 2, (ii) 3.

4.6 (i) 3, (ii) 2.

4.7 $\ker T = (0, 0, 1)$, dimension 1.
$\boldsymbol{T}(\boldsymbol{R}_3) = \boldsymbol{R}_2$, dimension 2.

4.8 Let $T \in L(\boldsymbol{R}_3, \boldsymbol{R}_2)$, $T(x_1, x_2, x_3) = (x_1, x_2)$. $\boldsymbol{x} = (1, -1, 2)$ and $\boldsymbol{y} = (2, -2, 5)$ are linearly independent in $\boldsymbol{R}_3$. But $T\boldsymbol{y} = 2T\boldsymbol{x}$, so that $T\boldsymbol{x}$ and $T\boldsymbol{y}$ are linearly dependent.

4.9 The kernel of S is the space of all polynomials of degree 0 (i.e., constants) and the image is the space of all polynomials of degree $< n$. These have dimension $1, n$ respectively and P has dimension $n + 1$.

5.1 $h(16, 5, 17, 0) + k(5, 9, 0, 17)$, where h and k are arbitrary $(1, -1, 2, -3)$.

5.2 $k(1, 1, -1)$, where k is arbitrary.

5.3 $x = y = z = 0$.

5.4 $x = 5h - k - 4$, $y = -7h + 9$, $z = h$, $w = k$, where h and k are arbitrary.

5.5 $x = 3h + 2k - 7$, $y = -2h - k + 6$, $z = h$, $w = k$, where h and k are arbitrary rational numbers.

5.6 $(1, 0, 0, 0) + h(-1, -1, 1, 0) + k(-1, 2, 0, 1)$, where h and k are arbitrary real numbers.

5.7 $x = k/3$, $y = h$, $z = h + 7k/3$, where h is arbitrary.

5.8 $\lambda = 2$, $x = -5k/3 + 7$, $y = 4k/3 - 3$, $z = k$, where k is arbitrary.

6.1 $\boldsymbol{A}$ has eigenvalues $+1, -1$ with corresponding eigenvectors $\{2, -1\}$, $\{1, -1\}$ respectively. $\boldsymbol{B}$ has eigenvalues 0, 3 with corresponding eigenvectors $\{1, -\sqrt{2}\}$, $\{\sqrt{2}, 1\}$ respectively. $\boldsymbol{C}$ has one eigenvalue 2 with eigenvector $\{1, 1\}$.

6.2 (i) Each eigenspace is of dimension 1. $\boldsymbol{A}$ is similar to

$$\begin{bmatrix} 1 & 0 \\ 0 & -1 \end{bmatrix}.$$

(ii) Each eigenspace is of dimension 1. $\boldsymbol{B}$ is similar to

$$\begin{bmatrix} 0 & 0 \\ 0 & 3 \end{bmatrix}.$$

(iii) Eigenspace is of dimension 1. $\boldsymbol{C}$ is not similar to a diagonal matrix.

6.3 $\boldsymbol{A}$ has eigenvalues $1, \sqrt{7}, -\sqrt{7}$ with corresponding eigenspaces $[(3, 1, 0)]$, $[(6 + 3\sqrt{7}, 3, -1 - 2\sqrt{7})]$, $[(6 - 3\sqrt{7}, 3, -1 + 2\sqrt{7})]$, respectively, each of dimension 1.

$\boldsymbol{A}$ is similar to

$$\begin{bmatrix} 1 & 0 & 0 \\ 0 & \sqrt{7} & 0 \\ 0 & 0 & -\sqrt{7} \end{bmatrix}.$$

B has eigenvalues $1, 3+\sqrt{2}, 3-\sqrt{2}$ with corresponding eigenspaces $[(0, 1, -1)], [(-1, \sqrt{2}, 1)], [(1, \sqrt{2}, -1)]$, each of dimension 1.

B is similar to

$$\begin{bmatrix} 1 & 0 & 0 \\ 0 & 3+\sqrt{2} & 0 \\ 0 & 0 & 3-\sqrt{2} \end{bmatrix}.$$

C has eigenvalues $1+i, 1-i, 1$ with corresponding eigenspaces $[(1, -1-i, 2)], [(0, 1, 0)], [(0, 0, 1)]$ respectively, each of dimension 1.

C is similar to

$$\begin{bmatrix} 1+i & 0 & 0 \\ 0 & 1-i & 0 \\ 0 & 0 & 1 \end{bmatrix}.$$

6.4 $\lambda^2 - 1 = 0$, $\boldsymbol{A}^7 = \boldsymbol{A}$, $\boldsymbol{A}^8 = \boldsymbol{I}$.

6.5 $\lambda^2 - 3\lambda = 0$.

6.6 $\boldsymbol{H} = \begin{bmatrix} 4 & 3 \\ -3 & -2 \end{bmatrix}$,

$\boldsymbol{H}^{-1}\boldsymbol{A}\boldsymbol{H} = \begin{bmatrix} 2 & 0 \\ 0 & -1 \end{bmatrix}$,

$\boldsymbol{A}^{12} = \begin{bmatrix} 9-2^{15} & 12-3\cdot 2^{14} \\ -6+3\cdot 2^{13} & 9\cdot 2^{12}-8 \end{bmatrix}$.

6.8 Eigenvalues of **A** are 6 and -2.

7.4 E.g., $M^{\perp} = [(1, -1, 1), (0, 1, 1)]$.
$\boldsymbol{p} = (-1/3, -1/6, 1/6)$,
$\boldsymbol{q} = (4/3, 1/6, 17/6)$.

7.5 $1 - 6t + 6t^2$.

7.7 $\lambda = -(\boldsymbol{u}, \boldsymbol{v})/\|\boldsymbol{v}\|^2$.

7.8 E.g.,
$M = [(1, 2, 2, 0), (-4, -5, 7, 12), (40, 11, -31, 36)]$.
$M^{\perp} = [(8, -8, 4, -3)]$, $(-3, -3, 9, 12)$.

7.9 $M^{\perp} = [(2(1+i), 1, -2i), (7/3 - 4i/3, i/3, -1/3)]$.

8.3 $\begin{bmatrix} 5/13 & 12/13 & 0 \\ 12/13 & -5/13 & 0 \\ 0 & 0 & 1 \end{bmatrix}$.

8.4 $\begin{bmatrix} (1+i)/2 & (1-i)/2 \\ (1+i)/2 & (-1+i)/2 \end{bmatrix}$.

8.5 C_1.

8.7 $\dfrac{1}{1+a^2}\begin{bmatrix} 1-a^2 & -2a \\ 2a & 1-a^2 \end{bmatrix}$.

8.8 $\boldsymbol{B} = \begin{bmatrix} 0 & a & 0 \\ -a & 0 & a \\ 0 & -a & 0 \end{bmatrix}$.

8.9 $\boldsymbol{A} = \begin{bmatrix} 0 & -i & 0 \\ -i & 0 & 0 \\ 0 & 0 & 1 \end{bmatrix}$.

9.1 (a) Yes. (b) No. (c) No., e.g., (1, 1, 3).

9.3 (a) $r = 2$, $s = 2$.
(b) $r = 1$, $s = 1$.
(c) $r = 3$, $s = -1$.

9.4 (a) $\begin{bmatrix} 3 & 2 \\ -2 & 3 \end{bmatrix}$,

(b) $\begin{bmatrix} 1 & 2 & 2 \\ 0 & 5 & -5 \\ -2 & 1 & 1 \end{bmatrix}$.

9.5 (a) $\sqrt{3}x_1 = \sqrt{2}y_1 + y_2$,
$\sqrt{3}x_2 = -y_1 + \sqrt{2}y_2$.
(b) $7x_1 = 2y_1 - 3y_2 + 6y_3$,
$7x_2 = 3y_1 + 6y_2 + 2y_3$,
$7x_3 = -6y_1 + 2y_2 + 3y_3$.
(c) $3\sqrt{5}x_1 = -\sqrt{5}y_1 + 6y_2 + 2y_3$,
$3\sqrt{5}x_2 = 2\sqrt{5}y_1 + 3y_2 - 4y_3$,
$3\sqrt{5}x_3 = 2\sqrt{5}y_1 + 5y_3$.

9.6 $\sqrt{6}x_1 = (1+i)(\sqrt{2}y_1 + y_2)$,
$\sqrt{3}x_2 = (-y_1 + \sqrt{2}y_2)$.

9.7 3.

10.1 $x_1 = 5y_1 + 3y_2$, $x_2 = 2y_1 + y_2$ gives $y_1^2 + y_2^2$, $-3y_1^2 + 2y_2^2$, respectively.

10.2 $x_1 = y_1 + y_2$,
$x_2 = -4y_1 - 2y_2 + y_3$,
$x_3 = 2y_1 + 2y_2 - y_3$,
gives $16y_1^2 - 4y_2^2 + 2y_3^2$,
$16y_1^2 + 4y_2^2 + y_3^2$, respectively.

10.3 $x_1 = y_1 + 2y_2 + 2y_3$,
$x_2 = -2y_1 - 4y_2 + 2y_3$,
$x_3 = -2y_1 + 5y_2 - y_3$ gives
$81y_1^2 + 324y_2^2 + 36y_3^2$,
$81y_1^2 - 324y_2^2 + 72y_3^2$,
respectively.

10.4 $x_1 = -2y_1 + y_2 - 4y_3$,
$x_2 = y_1 + y_2 + y_3$,
$x_3 = y_1 + 5y_3$ gives
$-4y_1^2 + 5y_2^2 + 20y_3^2$,
$4y_1^2 + 5y_2^2 + 20y_3^2$, respectively.

10.6 $2\pi\sqrt{\left(\dfrac{2a}{g}\right)}$, $2\pi\sqrt{\left(\dfrac{2a}{15g}\right)}$.

Index